THE SCIENCE OF
SOUND RECORDING

JAY KADIS

ELSEVIER

Amsterdam • Boston • Heidelberg • London
New York • Oxford • Paris • San Diego
San Francisco • Singapore • Sydney • Tokyo

Focal Press is an imprint of Elsevier

Focal Press

Focal Press is an imprint of Elsevier
225 Wyman Street, Waltham, MA 02451, USA
The Boulevard, Langford Lane, Kidlington, Oxford, OX5 1GB, UK

Notices
Knowledge and best practice in this field are constantly changing. As new research and experience broaden our understanding, changes in research methods, professional practices, or medical treatment may become necessary.

Practitioners and researchers must always rely on their own experience and knowledge in evaluating and using any information, methods, compounds, or experiments described herein. In using such information or methods they should be mindful of their own safety and the safety of others, including parties for whom they have a professional responsibility.

To the fullest extent of the law, neither the Publisher nor the authors, contributors, or editors, assume any liability for any injury and/or damage to persons or property as a matter of products liability, negligence or otherwise, or from any use or operation of any methods, products, instructions, or ideas contained in the material herein.

Library of Congress Cataloging-in-Publication Data
Application submitted

British Library Cataloguing-in-Publication Data
A catalogue record for this book is available from the British Library.

ISBN: 978-0-240-82154-2

For information on all Focal Press publications
visit our website at www.elsevierdirect.com

12 13 14 15 16   5 4 3 2 1

Printed in the United States of America

Working together to grow
libraries in developing countries

www.elsevier.com | www.bookaid.org | www.sabre.org

ELSEVIER   BOOK AID International   Sabre Foundation

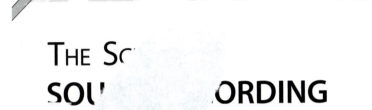

THE S...
SOU... ...ORDING

CONTENTS

INTRODUCTION

It is hard to believe that we have been able to record sounds for only a little over one hundred years. Those born after the digital revolution take the ability to make sound recordings entirely for granted. The widespread availability of inexpensive, high-quality recording equipment has greatly increased the number of audio recording practitioners. Although it is easy to get started, many soon discover that attaining the quality of recordings they expect is not as simple as sticking a microphone in front of an instrument and pressing the record button.

Many beginning audio devotees fail to consider the scientific principles behind the techniques professionals use to create the recordings they wish to duplicate. Long internships − once a staple of entry into the recording industry − are vanishing, while more beginners simply start on their own with little or no instruction. Once the difficulty of making great recordings becomes evident, those serious about a career in recording seek out more information. Many schools promise to make the student into a world-class engineer in a few simple lessons, often for a hefty tuition. Simply learning the current techniques without an understanding of why these particular approaches work as they do limits one's ability to deal with the challenges presented by new recording situations and with the continuing rapid development of recording hardware and software.

Science is a framework used to organize our observations of how the world works. It makes use of mathematics to quantify the behavior of physical objects. It relies on repeatable experiments to verify the theories we formulate about how things work. Although science is applicable to any endeavor that involves physical objects, it is often ignored when we try to figure out how something works. The complexity of mathematics can scare us away from its fundamental usefulness in analyzing the processes we wish to understand. Even without being able to solve differential equations, we can derive a useful understanding of how a system behaves from how the equations are set up and what variables are included. The goal of this book is to show how science explains the techniques we have at our disposal for making and manipulating sound recordings.

Though science is not a static collection of observations, it is a stable source of knowledge. As new discoveries are made, previously accepted ideas are altered to fit both the old and new information. Fortunately, most of the science applied to recording technology is old enough to have been time-tested and proven. It is well worth the effort to learn the scientific principles on which sound recording relies. Areas such as acoustics and electronics are well described and considered settled. Psychoacoustics, on the other hand, leaves much to be explained about how we translate physical air vibrations into musical sound that excites the emotions. The application of the scientific method in our quest to understand the intricacies of recording brings us ever closer to mastering the incredible power we have available to create any sound environment we can imagine.

We will start with a review of the mathematics used in the scientific explanations of acoustics and electronics. The science of measurement is also important because microphones essentially make a measurement of air pressure or velocity that becomes our recorded signal. The ability to quantify physical behaviors is central to building a structured understanding of the systems we use to record sound. The physics of mechanical systems and gases helps understand how microphones transfer sound wave information into electric signals. We also want to understand how we perceive sound, as our ability to hear determines what we must preserve in the processes involved in recording and playing back sound recordings. Of course, sound is at the center of the recording process, and understanding its behavior will begin to explain why microphone placement can be such a tricky proposition. The tools available to process the electronic representation of sounds are at the heart of studio practices. We will look at the passive and active electronic devices, often considered "black boxes," that make up our devices. Then we will look at the audio processing devices themselves.

Digital recording is the popular technology of the day, but the analog tape recorder was the dominant recording technology throughout the later 20th century and is still popular with rock music studios. Analog recorders, once the mainstay of the recording business, are being discovered by a new generation that appreciates the sound of tape enough to master the care and maintenance of these machines. New magnetic tape is being designed and manufactured to meet the continuing demand. Learning how a mature technology dealt with the limitations of the medium is enlightening.

Digital recording systems have been refined from the days of PCM (pulse-code modulation) to the video converters that began the digital audio age. Digital data is significantly different from the original analog signals that are produced by microphones, so examining the theories underlying digital audio will explain the limitations of analog-to–digital conversion as well as reveal the immense power of digital signal processing. The final chapter involves using all the technical information to accomplish the artistic goals of mixing sounds into the product: recordings for others to enjoy.

This book is not intended to give the reader a mastery of science but rather to reveal how scientific principles apply to the processes we use every day without a second thought. When we understand more of how the systems work, we become better at using them. The reader will find recommended readings that go further into the science of selected subjects to continue the quest we are only beginning here. One of the great aspects of sound recording is that one may enjoy learning the craft for a lifetime. May this book set you on that path.

CHAPTER ONE

Mathematics and the Measurement of Sound

Contents

Sound is inherently a transient phenomenon. Natural sounds fade quickly, and we can't just bottle up the air as it vibrates with sound waves and expect to hear the sounds again when we open the bottle. Although this behavior keeps the world from becoming a total cacophony, it also complicates the process of recording sounds. We need a method of converting such air vibrations into a form of energy that we can preserve over time. Metrology, the science of measurement, is a most important consideration in the recording of sound. To record and manipulate audio signals, we must first measure characteristics of the sound – generally pressure amplitude as a function of time – and in some manner preserve those measurements. Analog magnetic tape stores a continuous magnetic pattern representation of the measured signal. Digital audio systems store the signal as a list of discrete measurements at regular times that can then be treated like any other computer data. Because these measurements are a function of time, we must have a way of measuring time as well. Once we have a reliable record of time and the signal amplitude is stored, we can manipulate the signal electronically as analog audio or numerically as digital audio and then make changes to suit our artistic desires; we can process and mix sounds into sonic landscapes that never before existed. Appreciating that the quality of a recording depends on how well we are able to make a measurement is an important first step in the study of sound recording.

The Science of Sound Recording
ISBN 978-0-240-82154-2

A measurement is an attempt to determine the value of a quantity using some form of calibrated tool or procedure. The number we obtain from our measurement should be exactly equal to the true value of the quantity under measure, a quality we refer to as **accuracy**. The tools available will ultimately determine the accuracy of measurements because there is always a limit to their resolution – that is, their ability to distinguish between two close values. When dealing with calculations, we must also be careful that we do not generate a false increase in accuracy simply through mathematical processing of the data. For example, if we make a measurement accurate to two decimal places, multiplying that value on a computer may result in a 15-decimal-place answer, but it is still accurate only to the two decimal places of our original measurement. A second term related to measurement quality is **precision**, a measure of how closely that measurement can be replicated in repeated attempts. These two terms are sometimes used synonymously but in fact have slightly different meanings.

MATHEMATICS

To study the recording process in an organized way, we should first familiarize ourselves with the tools involved in the investigation. We are interested in the processes that allow the permanent storage of information contained in the transient air pressure variations we call *sound*. Because sound recording involves acoustics, mechanics, electronics, magnetism, and ultimately physiology and psychology, we need to employ the tools of science. Because we are concerned initially with measurements of the behavior of air pressures, electronic signals, and magnetism, physics is the branch of science that will help develop our understanding of these systems, and mathematics is the descriptive language used by physicists. It turns out that similar mathematical descriptions apply to electronics, acoustics, and mechanics, so understanding what equations mean will help us explain the fundamental principles of sound recording.

When we want to see how two quantities are related, we can plot a graph comparing pairs of values termed **variables**. The variable we set by choosing its value is called the **independent variable**; it determines the placement of the point along the x-axis. The variable we measure is called the **dependent variable** because its value depends on the value of the independent variable we choose. The dependent variable determines the y-axis placement of the point. For example, we might wonder how the amplitude of a sound system varies with frequency. We measure the dependent variable (amplitude) at a number of frequencies we choose (the independent variable) and plot amplitude against frequency. The shape of the resulting curve shows us the function that describes the relationship between the two properties, in this case the frequency response of the system. Sometimes there is no obvious relationship, and sometimes the relationship between the variables can be described by an equation, called a **function**, detailing how they are related mathematically.

When our measurements fall along a straight line, we call the relationship **linear**. Linearity is a requirement of most audio processing systems. For instance, we want the output of an audio amplifier to be a linearly scaled version of the input. The gain of an amplifier is the ratio of its output amplitude to its input, so a gain greater than 1 indicates amplification; a gain less than 1 means attenuation. In Figure 1-1 (left), the slope of the line plotting output voltage (y-axis) against input voltage (x-axis) shows the gain. For a linear system, this value is a constant; in our example, the output is always half ($\times$ 0.5) the input level regardless of the input voltage. Other relationships, such as the intensity of a sound as a function of the distance from the source, are not linear (Figure 1-1, right.) Nonlinear relationships can cause **distortion**, a change in the shape of a signal. Comparing linear and nonlinear systems may help clarify the difference.

As long as the dependent variable depends only on constant multiples of the independent variable, there is a linear relationship between them. Linear audio processes include **mixing**, in which signals are added together, and **amplification**, in which a signal is multiplied by a constant, the gain factor. Nonlinear audio relationships include modulations, with one variable

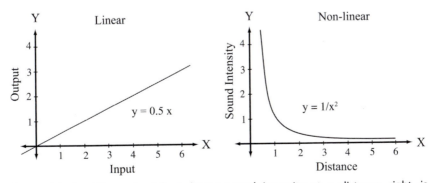

Figure 1-1 Amplifier gain, left, is linear; sound intensity at a distance, right, is a nonlinear function.

multiplied by another (as in AM and FM radio) and exponential or logarithmic relationships found in dynamic range compressors and expanders. Nonlinear systems can create distortion of the original signal by introducing harmonics (multiples) of signal frequency components that were not present in the original signal source.

SINUSOIDS

Audio signals, whether measured as sound pressure levels or electronic voltages, are often represented as sine waves. Music – in fact, all sounds – may be described using combinations of sine waves with varying amplitudes, frequencies, and phase relationships. The **Fourier series** equation is commonly used to describe the combination of sinusoidal components that constitute a sound. Although real musical signals are more complex, we often use sinusoidal test signals when measuring audio systems. **Sine waves** themselves are actually functions (see Figure 1-2).

In Figure 1-2, the signal voltage y equals a maximum value A (A $= 1$ in Figure 1-2) multiplied by the sine of the variable x, with x values ranging from 0 to 2π to generate one complete cycle. The sine function can be generated by the rotation of a point on a circle with a radius of 1, called the **unit circle** (Figure 1-3). If the point starts at the 3 o'clock position and rotates counterclockwise along the circle circumference, its x and y values generate a sine wave. (When the point starts at the maximum $y = 1$ (12 o'clock) instead of $y = 0$, it generates the cosine function.) At any angle of rotation, the height of the sine wave along the y-axis will be the sine of the angle and the x-position will be the angle of rotation in radians. Though we are familiar with degrees as the measure of angles, where 360° describes a full

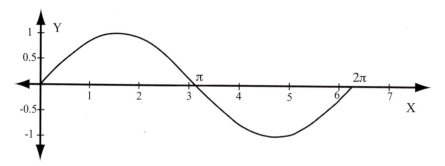

Figure 1-2 A graph of the function y $=$ A sin(x).

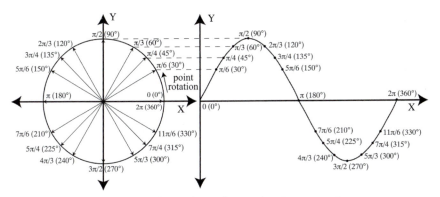

Figure 1-3 The unit circle and its relationship to the sine wave. As we rotate counterclockwise along the unit circle, the y-axis value moves up and down as a function of the angle of rotation. The angle gives the corresponding x-axis value in radians.

circle, mathematics uses the radian as the angular measure. There are 2π radians in a full circle and therefore in a complete sine wave cycle. For a point (x,y) on the sine function, the sine is the y value; the cosine is the x value.

The **phase** of a sine wave is a measure of the displacement from the origin (0,0), along the x-axis, of its beginning, where it crosses zero on its way up. When beginning at the origin, the phase is taken to be zero. The amount of rotation along the unit circle required to turn one sine into the other is the **relative phase angle** between them. As we add sine waves together to make a complex signal, the relative starting points of the various sine components determine the phase relationship between them. If two sine waves of the same frequency are exactly in phase, they add constructively. If they are $180°$ (π radians) out of phase, they cancel completely – provided they are of the same amplitude. We will see examples of such cancelations when we consider room acoustics and microphone polar patterns. The combination of constituent sine waves determines the **timbre**, or sonic texture, of the sound. Electronic circuits known as **filters** alter the timbre by changing the balance of sinusoidal components. Filtering often alters the relative phases of the different frequency sine components of a signal in rather complex ways. We will explain in later chapters how both analog and digital filters affect the relative positions in time of a signal's component sinusoids as a function of frequency.

To keep track of our work, we are interested in displaying information about our audio signals graphically. By far the most common audio measurement is the signal amplitude. We need to create a meaningful display of the amplitude measurement that is easily interpreted and relatively inexpensive to duplicate, especially if we're dealing with dozens of separate signals, as is often the case with multitrack recording. Early audio recording equipment solved this problem by using mechanical meters with a moving needle that were carefully calibrated so that all such meters had the same mechanical characteristics. More recently, we have seen the popularity of light-emitting diode (LED) ladder displays. Each of these graphical indicators must conform to a set of rules about how we measure and display signal levels in order to be unambiguous. Because there are alternative methods of measuring the amplitude of signals, we should know which of these is used in our graphical display.

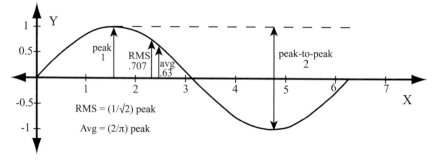

Figure 1-4 Sine wave amplitude measures.

The amplitude of a sine wave can be measured in different ways (Figure 1-4). The **peak-to-peak** measurement will tell us the swing between maximum and minimum values of our signal, but this measurement may not directly correspond to the apparent loudness of that signal when it is not a simple sine wave. The method that most closely approximates how we experience the loudness of a signal is the **root-mean-squared (RMS) measure**, in which we analyze the signal level over some time window and compute the square root of the mean of the squared values of the measured levels (see Equation 1-5, below). This is a complicated measurement to make electronically, so a simpler method is often used: the **average** value. This value can be computed with simple analog circuitry, but it is a slightly less accurate measure of the perceived loudness of the signal. One problem with both of these approaches is that they average the signal – a process that will fail to track the maximum value of the signal. Because the maximum value determines how much signal amplitude the audio system must be able to accommodate, we are in danger of overloading the electronics if we simply measure the average level and ignore the peaks.

Multiply by to Convert from Below to >>	Peak-to-Peak	Peak	Average	RMS
Peak-to-Peak	1	0.5	0.3185	0.3535
Peak	2	1	0.637	0.707
Average	0.785	1.57	1	1.11
RMS	2.828	1.414	0.9	1

Figure 1-5 Sine wave amplitude measurement conversion factors.

Therefore, very fast displays have evolved to augment the loudness-oriented displays. These peak-oriented displays may be as simple as an LED that flashes as the signal level approaches the limits of the electronic system.

For sine waves, there is a simple conversion factor between these different measurements (see Figure 1-5). We notice that for sinusoidal signals, the average level is 63.7% of the peak value and RMS is 70.7%. Unfortunately, for actual complex audio signals, this simple relationship does not hold, so meters using different methods may not agree. In the studio, we often use equipment calibrated to different standards of measurement, so we need to understand how they relate if we want to get the best performance from the system. The sine wave test signals often used for measuring circuit performance are **steady state**, meaning that they do not change in amplitude or frequency over time. It is this characteristic that allows the simple relationship between peak, average, and RMS measurements. Real audio signals are continuously changing in both amplitude and frequency content. Short, rapid changes in the signal are known as **transients**, which cause circuits to behave differently from steady-state signals when the circuits contain capacitors or inductors, as these elements are sensitive to the rate of change of the applied voltage or current. The difference between steady-state and non-steady-state signals also becomes important when evaluating circuits that deal with dynamic range–processing devices, for example.

Figure 1-6 shows a typical audio signal. The peaks exceed the average level by many decibels. The ratio of peak amplitude to RMS amplitude is known as the **crest factor**. A type of meter designed to measure the peak amplitude of a signal is known as a **peak program meter (PPM)**. Many digital meters display both RMS and PPM measurements simultaneously.

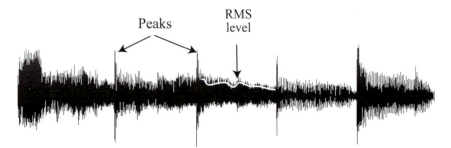

Figure 1-6 A typical music waveform. The RMS level corresponds to the perceived loudness; peaks may exceed the average level by 10 decibels or more.

LOGARITHMS AND EXPONENTS

The range of values we encounter when measuring amplitude and frequency is enormous. The "ideal" human can hear frequencies from 20 Hz to 20 kHz (few people actually can) and hear sound levels from silence to painfully loud – about six orders of magnitude in sound pressure amplitude. Working with numbers over this large range is inconvenient, so we use **logarithms**. The logarithm function is graphed in Figure 1-8. Logarithms are **exponents**: the power to which a base number must be raised to equal the number in question. For example, the $\log_{10}$ (log base 10) of 100 ($= 10^2$) is 2. In the audio world, we generally use $base_{10}$ (unless otherwise designated, "log" implies $base_{10}$); however, many physical processes are described by the so-called natural logarithm ln (base e $= 2.718\dots$ – don't ask) and digital signal processing is conducted ultimately in binary, or base $_2$. By using logarithmic measures, we avoid lots of zeros and minimize the number of digits.

There are properties of logs that simplify their use. Any base to the zero power equals 1. The power ½ is the square root. Negative exponents yield numbers less than 1. Positive and negative exponents are reciprocals, for example $10^2 = 100$, $10^{-2} = 1/100$. Positive whole-number exponents ($base_{10}$) correspond to the number of zeros following the 1. Logarithms exist only for numbers greater than zero. The exponential notation used for large and small numbers in the metric system has been standardized with prefixes denoting the exponent. These are listed in Figure 1-7. **Scientific notation** makes use of exponents. The expression 6.02×10^{23} (the number of "things" in one mole) is expressed in scientific notation.

The primary audio application of the logarithm is the decibel. Because signal amplitude measurements must be made over many orders of

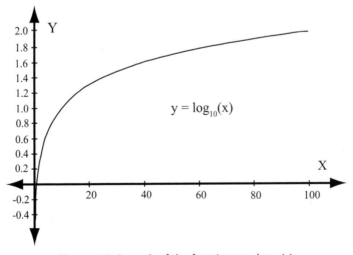

x	10^x
-2	0.01
-1	0.1
-0.5	0.303
0	1
.303	2
.5	3.16
.603	4
.903	8
1	10
2	100
3	1000

base → 10^x ← exponent

10^{-1}	deci- (d-)	10^{24}	yotta- (Y-)
10^{-2}	centi- (c-)	10^{21}	zetta- (Z-)
10^{-3}	milli- (m-)	10^{18}	exa- (E-)
10^{-6}	micro- (μ-)	10^{15}	peta- (P-)
10^{-9}	nano- (n-)	10^{12}	tera- (T-)
10^{-12}	pico- (p-)	10^{9}	giga- (G-)
10^{-15}	femto- (f-)	10^{6}	mega- (M-)
10^{-18}	atto- (a-)	10^{3}	kilo- (k-)
10^{-21}	zepto- (z-)	10^{2}	hecto- (h-)
10^{-24}	yocto- (y-)		

Figure 1-7 Exponents in powers of ten. The metric prefixes are used in scientific notation.

$y = \log_{10}(x)$

Figure 1-8 A graph of the function $y = \log_{10}(x)$.

Measurement Quantity	Reference Level
Sound pressure level	0 dB SPL = 20 μ pascal
Voltage	0 dBV = 1 volt
Voltage	0 dBu (or dBv) = 0.775 volt
Power	0 dBm = 0.001 watt

Figure 1-9 Amplitude reference levels.

magnitude, the bel ($\log_{10}[\text{power}_1/\text{power}_{ref}]$) was adopted as a unit of measure by the early telecommunications industry. Though appropriate in telegraphy where signal power transmission was important, the unit was too large to be convenient for electronic audio circuits and the **decibel** (1/10th bel, abbreviated dB) is now used. An important attribute of the decibel is that it isn't an absolute measure, but rather a ratio. It is used to describe how much larger or smaller a sound level or signal amplitude is than a standard reference level. Reference levels are chosen according to the application: they may represent the quietest sound we can perceive (dB sound pressure level), the maximum signal a system can produce (dB full scale), the recommended input level a device is designed to see (dBv), a power level (dBm), or whatever we choose (VU – volume units). Each of these assumes a different reference quantity (Figure 1-9). Power ratios are $10\log(x)$; voltage, current, and sound pressure levels are $20\log(x)$. Appendix 1 shows some examples of calculations using decibels.

$$decibel = 10 \log_{10} \frac{power_{measured}}{power_{reference}} = 20 \log_{10} \frac{voltage_{measured}}{voltage_{reference}} \qquad 1\text{-}1$$

VECTORS

In dealing with physical systems, we are often describing actions involving **forces**. A force has two components: a magnitude and a direction or phase. In order to mathematically describe a force, we must use **vector** math. An example of a vector measurement we commonly encounter is wind velocity: it has a magnitude (speed) and a direction. (Measures of magnitude without a direction are called **scalars** – temperature or mass, for example.) We can draw a vector as an arrow of length proportional to its magnitude and pointing in the direction of the action. If we wish to

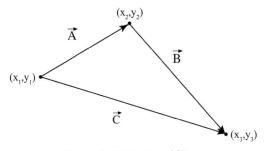

Figure 1-10 Vector addition.

add or combine two forces, we must use vector math in order to find the resulting force.

Figure 1-10 shows how vectors add graphically. Vectors $\vec{A}$ and $\vec{B}$ combine to produce vector $\vec{C}$; in order to get the magnitude and phase or direction correct, we must use vector addition. Here the origin of the second vector $\vec{B}$ is placed so that it originates at the terminal point of the first vector $\vec{A}$. This behavior is possible because absolute vector position is arbitrary; that is, the action is the same no matter where in space it occurs. As using vector math simplifies computations for mechanical, acoustic, and electronic systems, you need to understand only that vector math is computed differently from scalar math.

POLAR COORDINATES

Another technique often used in describing magnitude and direction is **polar coordinates** (Figure 1-11). Instead of plotting vectors by giving (x,y) pairs, polar coordinates consist of a magnitude and an angle (r,θ). This technique simplifies tasks like describing microphone directional sensitivity patterns and sound radiation patterns. For example, in a microphone sensitivity pattern chart, the length of the line r indicates the microphone's sensitivity (output voltage for a given sound level input) and the angle θ indicates the direction from which the sound originates. In Chapter 6, we will use polar coordinates to describe the spatial sensitivity of microphones.

COMPLEX NUMBERS

Vectors can also be represented as **complex numbers** (Equation 1-2) numbers with real (x) and imaginary (yi) components. Though this is not

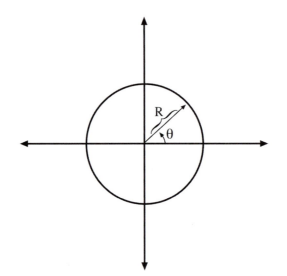

Figure 1-11 Polar coordinates.

our usual way of thinking about numbers, the use of imaginary numbers simplifies the analysis of acoustic and electronic systems. The square root of −1 is an imaginary number, because there is no real number that, when squared, generates a negative number. So we make some up: i and j. These are simply defined as the square root of −1, even though we don't come across such numbers in our daily activities (unless we're physicists or engineers). The real part of the number represents the magnitude of the vector, and the imaginary part determines the direction or phase, so we can use a single number to represent both components of the vector. Using complex numbers, vector computations become more convenient. The complex numbers used in electrical engineering are known as **phasors**. We will not need to use vector mathematics much in our quest for a conceptual scientific understanding of sound recording, but we need not be intimidated by vector math when we encounter it in technical papers. It is merely a way to simplify the task of analyzing the behavior of electronic and acoustic systems.

$$z = x + yi \qquad\qquad\qquad 1\text{-}2$$

Complex numbers are important in describing both analog and digital audio processes. Analog circuits are analyzed using complex numbers known as phasors (from phase vector), allowing an AC circuit to be analyzed in a simpler fashion, similar to the frequency-independent DC case. In digital signal processing, the Fourier transform used to analyze and process audio signals takes advantage of complex numbers. Regardless of whether we can grasp the idea of imaginary numbers, their use simplifies many analyses by providing a convenient mathematical method.

CALCULUS

Another branch of mathematics employed in many of the physical analyses we encounter in sound recording is **calculus**. Calculus is considered mystifying by many and is often avoided for that reason. Although algebra may be sufficient to describe simple physical systems, the application of calculus actually simplifies the analysis of many of the changing physical interactions we encounter in the processes of sound recording. Calculus is the mathematics of change. It consists of the **derivative**, the function that describes the rate of change of a function, and its inverse function the **integral**, which describes the area below the function and above the x-axis between two specified points.

Figure 1-12 shows the two basic forms of calculus. The derivative of function $f(x)$ is the slope at each point along the curve, which becomes a new function dy/dx. The shaded area under the curve is the integral, in this case from 2 to 4, of the function $f(x)$ (see Equation 1-3):

$$\int_{x=2}^{x=4} f(x)\,dx \qquad\qquad 1\text{-}3$$

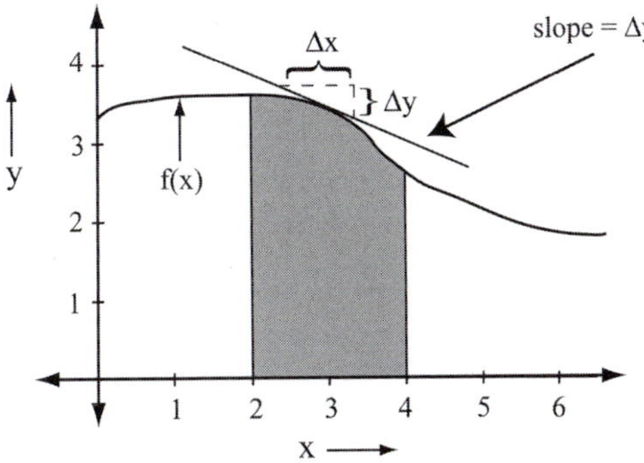

Figure 1-12 Calculus – derivatives and integrals.

In many cases, a function contains more than one variable. For instance, a moving object in space has variables for position along the x-, y-, and z-axes and one for time, $f(x,y,z,t)$. If we are interested in only one of these variables, a **partial differential** may be taken. The symbol for a partial derivative is ∂, so a partial derivative of $f(x,y,z,t)$ with respect to x would be $\partial f / \partial x$ where y, z, and t would be regarded as constants.

Integration and differentiation are inverse processes like multiplication and division or addition and subtraction. Although the ideas of calculus are essentially rather simple, when they are applied to real-world physical descriptions, the resulting equations can get quite complicated. Furthermore, many differential equations do not have exact solutions and must be evaluated using numerical methods. For our purposes, you need understand only what derivatives and integrals mean to see what many of the equations used to describe sound wave propagation or magnetic field interactions are telling us about the underlying physics.

STATISTICS

Frequently, the mathematical analyses we encounter describing the physics of sound recording deal with large numbers of individual events. For example, the behavior of air is ultimately the result of enormous numbers of individual molecular interactions: collisions between spinning diatomic gas molecules that transfer kinetic energy with each collision. When we attempt to understand why a sound system is perceived to sound a certain way, we must analyze the perceptions of many individuals in order to begin to explain what causes us to hear what we hear. These are examples of "population effects" that are best analyzed using the techniques of statistics.

Many people use the concept of the average in everyday life without knowing the mathematical definition. The average with which we are most familiar is technically the **arithmetic mean** ($\bar{x}$, Equation 1-4), the sum of all individual measurements divided by the number of events measured.

$$\bar{x} = \frac{x_1 + x_2 + \ldots + x_n}{n} \qquad 1\text{-}4$$

A very common measurement in sound recording that benefits from statistics is the average amplitude of a signal that is continuously changing. We measure the signal over time and create a statistic that describes its value over the period of measurement: the **root-mean-squared** or **RMS** value.

The RMS value is the square root of the arithmetic mean of the squares of the individual values (Equation 1-5).

$$\text{RMS} = \sqrt{\frac{x_1^2 + x_2^2 + \dots + x_n^2}{n}} \qquad\qquad 1\text{-}5$$

For RMS measurements of a continuous function, like a signal, the mathematical definition is more complicated but conceptually similar. The RMS measurement can be performed by dedicated integrated circuits or through digital computation. RMS measures the average power dissipated in a circuit when the voltage is changing over time. The arithmetic mean is most often used in statistics and is the most common statistic we encounter both in everyday life and in scientific experiments. The mean tells us the average value of the variable under consideration.

UNITS OF MEASURE

In order to apply mathematical analysis to our measurements, we need a standardized set of units of measure. Unfortunately, the quest to standardize a system of measurement has led to several competing systems in different parts of the world. Most of the world uses the metric system while the United States continues to use the older British system. The scientific community has adopted the metric system, which will be used here, although we will sometimes include British measurements as well because they are more commonly understood in the United States.

The metric system uses the meter as a unit of length, the kilogram as the unit of mass, and the second as the unit of time. For large measurements, the **MKS** (meter-kilogram-second) system is convenient; for smaller measurements, the **CGS** (centimeter-gram-second) system can be used. Scientists have adopted a system known as the International System of Units (**SI**) as their preferred system of measure. There are several units considered **base units**; that is, they are the fundamental unit of measure. They may be combined to generate **derived measures** that are more convenient to use. The derived measures are defined in terms of combinations of base units. The SI base units are listed in Figure 1-13.

Some frequently used SI and CGS units of measure are shown in Figure 1-14.

Quantity	SI Base Unit
distance	meter (m)
mass	kilogram (kg)
time	second (s)
temperature	kelvin (K)
electric current	ampere (A)
light intensity	candela (C)
number of elementary entities	mole (mol)

Figure 1-13 SI base units of measure.

Quantity	SI Unit of Measure	CGS Unit of Measure
distance	meter	centimeter
time	second	second
mass	kilogram	gram
force	newton (N)	dyne (= 10^{-5} N)
pressure	pascal (P)	barye (= 0.1 P)
work, energy	joule (J)	erg (= 10^{-7} J)
electric current	ampere (A)	ampere (A)
electric field strength	volt/meter	volt/centimeter
magnetic flux	weber (Wb)	maxwell (Mx= 10^{-8} Wb)
magnetic field strength	amperes/meter (A/m)	oersted (Oe= 79.6 A/m)

Figure 1-14 SI and CGS units of measure.

Derived SI units can be expressed as combinations of the base units. For example, the newton can be expressed as $m \cdot kg/s^2$; volts are $m^2 \cdot kg/s^3 \cdot A$. It is easy to see why we prefer to use the derived units.

CHAPTER TWO

Physics

Contents

When people think of sound recording, they usually think of microphones, wires, mixing boards, and tape recorders. Or perhaps they think of computers and plug-ins these days, but the basic foundation of the process of recording sound lies in the physics that describes the behavior of molecules and electromagnetic fields. The most critical elements of the recording process – the transduction of air pressure variations into proportional voltage changes and the storage of the converted information as magnetic or optical records (also known as analogue recording) – involve the science of **physics**. Physics is a branch of science that seeks to describe and quantify the behavior of matter and energy in the precise language of mathematics. Because the efficiency of the interaction of air molecules with a microphone transducer element determines how accurately we are able to record the different sound frequencies and how the microphone responds to sounds from different directions, we benefit from an understanding of the physics involved. Although the final decision about how well a microphone performs rests with our ears, understanding the physics involved helps us decide which microphone type and placement is likely to produce the desired result without listening to every microphone in every position possible.

The Science of Sound Recording
ISBN 978-0-240-82154-2

NEWTON'S LAWS OF MECHANICS

We begin our discussion of physics by considering the basic principles of Newtonian mechanics. Much of our modern understanding of the physical world begins with the 17th-century work of Sir Isaac Newton. Though everyone who has thrown a baseball innately knows how it will behave, the mathematical equations that predict the path it will take are less well known. Newton's three laws describe the basic behavior of matter and energy in simple mathematical terms.

To understand the physics of mechanical systems, we need to consider the basic properties of matter. **Mass** (kg) is the property we associate with the weight of an object; however, strictly speaking, that measurement is due only to our earthly frame of reference and the pull of gravity. Nonetheless, our everyday experience gives an intuitive understanding of what mass is: the more mass an object has, the heavier it feels and the more energy is required to move it. The more mass an object has, the greater its **inertia**: its ability to resist a change in movement. A related concept is **momentum** (kg·m/s): the product of mass and velocity. The momentum of an object measures how much energy it is able to provide when its motion is changed or stopped, that is, how much work it can do. **Displacement** (m) (Figure 2-1) is the distance and direction an object moves. **Velocity** (m/s) is the rate and direction of displacement. Its magnitude is the first derivative (the rate of change) of position with respect to time (dx/dt) for movement along the x-axis. (The first derivative

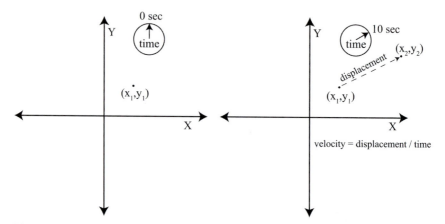

Figure 2-1 Displacement is the movement of an object from one point in space to another. Velocity is the rate of change of position.

of $f(x)$ is sometimes denoted as x'.) Because displacement, velocity, and momentum have both magnitudes and directions, they are vector quantities. (The magnitude of velocity is the scalar we call *speed*.) Changing velocity is **acceleration** (m/s^2), the first derivative of velocity (dv/dt) with respect to time and the second derivative of position (d^2v/dt^2) with respect to time. (The second derivative is the derivative of the first derivative.) **Force** describes a push or pull on an object, like the pull of gravity or the electric force of voltage on electrons. The more mass an object possesses, the more force must be applied to move it. The product of force times the distance moved is **work**, measured in joules (J $=$ N·m or m^2·kg/s^2). It is equivalent to the amount of **energy** transferred doing the work. The work done per unit time, the rate of energy transferred, is called **power** (watt (W) $=$ J/s). Although we intuitively know these terms, their formal definitions are less obvious.

Physical quantities may be classified based on whether their value depends on the size of the system or on the quantity of a material. For instance, density, pressure, and temperature do not depend on the amount substance involved. Mass, on the other hand, does depend on the amount of substance constituting an object. Properties that are independent of quantity are known as **intensive properties**; those that depend on quantity are **extensive properties**. When work is done, both an intensive property and an extensive property are involved. The intensive property provides the driving force and the extensive property responds to the force. In electricity, voltage is the intensive quantity and current is the extensive quantity. In mechanics, force is the intensive quantity and velocity is the extensive quantity. In acoustics, pressure is the intensive quantity and volume velocity is the extensive quantity. The rate of work done, or power, is the product of the intensive variable times the extensive variable.

Newton's First Law

Newton's First Law states that if no force is exerted on an object, its velocity cannot change, or, more commonly, "a body at rest tends to remain at rest." However, a moving object will not change its motion without an applied force, either. Although we understand intuitively what a force is, it is less clear how to formally describe it. Loosely, force is a push or pull on an object. We deal with forces when we consider how air movement propagates and pushes and pulls on the diaphragm of a microphone. Forces are also involved in electric circuits and in magnetic attraction and repulsion.

Newton's Second Law

Newton's Second Law states that the net force on an object is equal to the product of the mass and its acceleration, or in equation form (Equation 2-1):

$$\vec{F} = m\vec{a} \qquad\qquad 2\text{-}1$$

where $\vec{F}$ is the force vector (N); m is the mass (kg); and $\vec{a}$ is the acceleration (m/s^2), the rate of change in velocity. Note that force and acceleration both have a direction of action as well as a magnitude and are therefore represented as vectors. This simple relationship describes a surprising percentage of physical interactions.

Newton's Third Law

Newton's Third Law states that when two objects interact, the forces exerted by the objects on each other are equal in magnitude and opposite in direction, or, "for every action there is an equal and opposite reaction." It applies to billiard balls bouncing off one another, much like molecules behave on a smaller scale. Starting with Newton's laws, we are able to derive many of the equations used to describe the physics involved in recording sound.

THERMODYNAMICS

Thermodynamics is the study of the flow of energy. Physical systems that involve forces and movement, where work is done, involve the flow of energy. As sound waves propagate through the air, energy exchange is involved as the wave moves outward. The basic laws of thermodynamics can be used along with Newton's laws to derive most of the elementary relationships in the study of sound.

The **first law of thermodynamics** states that the *internal energy* of a system is conserved. The internal energy is the sum of the system's heat and work content. In an open system where heat or work can be transferred into or out of the system, the change in internal energy must equal the energy transferred in or out.

ELECTROMAGNETISM

Because we use electronic and magnetic representations of sound waves, we also need to understand **electromagnetism**. Electric current and magnetism are two inseparable entities. Electric current flow produces a magnetic field while moving a conductor through a stationary magnetic field produces a current in the conductor, a process called **induction**. Electromagnetic energy radiates as an electromagnetic **wave**. Waves are disturbances that travel through space, generally accompanied by a transfer of energy. Electromagnetic radiation consists of two transverse waves, one electric and one magnetic, oriented perpendicular to each other (Figure 2-2). In transverse waves, the wave motion is perpendicular to the directions of travel; in longitudinal waves, the movement is in the same direction as propagation. Electromagnetic wave classification is based on frequency, from low-frequency radio waves through light and ultimately high-frequency x-rays and gamma rays. The energy they contain is proportional to their frequency. Note that sound waves are not electromagnetic waves.

Electronic circuits are analogous to mechanical systems. Voltage is the equivalent of mechanical force, pushing charges through a circuit. An inductor acts as a mass, storing energy in a magnetic field; capacitors behave like springs, storing energy in an electric field. Resistors act like friction, simply dissipating energy as heat while opposing the flow of charge. Magnetism exerts a force on susceptible particles, much like a voltage does on an electrically charged particle. Magnetic and electric forces have a direction as well as a magnitude and are therefore represented by vectors. The distribution in space of electric and magnetic forces is described by a collection of vectors called a field, with a magnitude and direction for every point in space. The representation of magnetic fields is an arrangement of lines of force emanating from one pole (north or south pole for magnetism and positive or negative

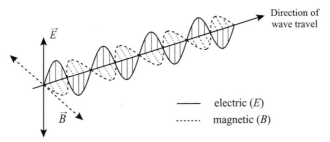

Direction of wave travel

$\vec{E}$

$\vec{B}$

—— electric (E)
······· magnetic (B)

Figure 2-2 An electromagnetic wave. The electric vector is perpendicular to the magnetic vector component.

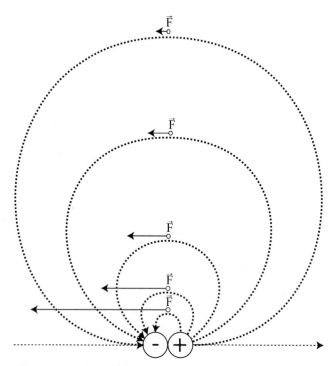

Figure 2-3 Electric dipole lines of force. The longer the path, the lower the force at each point on the path.

charge for electricity) and terminating at the opposite pole. The greater the path length, the weaker the force at points along that path (Figure 2-3). (This relation makes sense if you consider that the overall strength of the force between the poles is constant and the distance over which the force is exerted grows as the path lengthens. Therefore, the strength of the force at a point along the path decreases as the path length increases.)

WORK AND ENERGY

When a force changes the movement of an object, energy is transferred to the object and work is done. Energy may take one of two forms: **potential** or **kinetic**. An object at rest may possess potential energy, but its kinetic energy is zero. A moving object has kinetic energy and may also have potential energy. The amount of kinetic energy in a physical system is:

$$K = \frac{1}{2}mv^2 \qquad\qquad 2\text{-}2$$

where K is the kinetic energy (J), m is the mass (kg), and v is the velocity (m/s). Potential energy represents the ability to do work even if no work is currently being done: a ball at the top of a hill, for example. Mechanical systems allow for the back and forth interchange of potential and kinetic energy; however, the total amount of energy in a closed system remains constant. In the case of gravitational potential energy, the relationship is represented by the equation:

$$U(\gamma) = mgh \qquad\qquad 2\text{-}3$$

where $U(\gamma)$ is the potential energy at a height of h (m), m is mass (kg), and g is the acceleration due to gravity (9.8 m/s^2). There are other forms of potential energy – tension in a spring, for example. Voltage, also called potential and electromotive force (EMF), is a form of potential energy.

Potential and kinetic energy drive the behavior of mechanical, electric, and acoustic systems in similar ways. Figure 2-4 shows the contributions of these forms of energy to the functioning of the physical systems.

Kinetic energy is generated by motion, whether it is the motion of charges in an electric circuit, a moving solid mass, or a moving column of air molecules. Potential energy may be stored as an electric field in a capacitor, as tension in a stretched spring, or as pressure that is increasing or decreasing in a closed volume of air. Similar equations describe these manifestations of energy action.

Physical System	Dissipative Property	Kinetic Property	Potential Property
Electrical	Resistance (Energy dissipated as heat)	Inductance (Current flow stores energy in magnetic field)	Capacitance (Energy stored in electric field)
Mechanical	Mechanical Resistance (Friction)	Mass (Momentum)	Compliance (Energy stored in spring tension)
Acoustical	Acoustic Resistance (Viscosity of air)	Inertance (Inertia of air column)	Acoustic Capacitance (Energy stored pressure changes)

Figure 2-4 Properties of mechanical, electrical, and acoustical systems.

RESONANCE AND HARMONIC MOTION

Physical systems combining both potential and kinetic energy may exhibit the interesting and useful phenomenon of **resonance**. Resonance is a special case of **harmonic motion**. Harmonic motion occurs when the displacement of a mass results in a restoring force that is proportional to the displacement distance. The example of a weight hanging on a spring illustrates harmonic motion in a familiar way. When the weight is lifted and dropped, the potential energy of the raised position is converted into kinetic energy as the weight falls. The spring is stretched and begins exerting an upward restoring force on the falling object, converting the energy of motion into potential energy as tension in the spring. Eventually, the potential energy of spring tension overcomes the kinetic energy of the falling weight and it stops its fall and begins moving upward, converting the spring tension back into kinetic energy as upward movement. This interchange of energy continues until it is dissipated as heat in the stretched spring and by the slight but finite resistance of air on the moving weight. The resonance is damped as the energy is dissipated and oscillation ulti-mately stops. Similar resonant exchanges of energy occur in electronic circuits and in acoustics. It is resonance that shapes the sound of musical instruments and room reverberations.

THE WAVE EQUATION

Waves are important phenomena in physics. Waves describe the propagation of electromagnetic energy, sound, and the vibration of a stretched string. General wave action is described by the **wave equation**. Mechanical waves like sound and water waves travel through an elastic medium and obey Newton's laws. Electromagnetic waves propagate without a medium. Waves are classified by the direction of motion relative to the direction of propagation; in transverse waves like string vibration, the motion is perpendicular to the direction of propagation, and longitudinal waves like sound waves are characterized by motion in the same direction as propagation. Both types of waves are **traveling waves** because they propagate. Stationary or standing waves are also possible.

A general description of wave motion is given by the partial differential Equation 2-4, where v is the speed of propagation:

$$\frac{\partial^2 y}{\partial x^2} = \frac{1}{v^2} \frac{\partial^2 y}{\partial t^2}$$

2-4

For a one-dimensional wave, like a string vibration, the equation describes the y value as a function of x position, time and the speed of propagation. To solve for the positions, the equation must be integrated twice, a process that results in two constants of integration. Equation 2-5 is a general solution to the one-dimension wave equation:

$$y(x, t) = f_1(x \pm vt) + f_2(x \pm vt) \qquad \text{2-5}$$

By limiting the oscillation to a single frequency, Equation 2-5 yields Equation 2-6:

$$y(x, t) = A \sin[k(x - vt)] + B \cos[k(x - vt)] \qquad \text{2-6}$$

In Equation 2-6, A and B are the maximum amplitudes of the two sinusoids. Their values can be established by using known boundary conditions, like their values at $x = 0$, $t = 0$.

$$y(x, t) = y_{max} \sin[k(x - vt) + \phi_0] \qquad \text{2-7}$$

In Equation 2-7, $y_{max} = \sqrt{A^2 + B^2}$ and $\phi_0 = \tan(B/A)$, from Equation 2-6. The general wave equation is extended to three dimensions in space and is used frequently in describing physical systems involving wave propagation.

SUPERPOSITION

The principle of **superposition** describes the process of combining two linear functions. If two functions are linear, they may be combined by addition. This principle is used to describe the summation of the responses to two or more stimuli in linear physical systems, such as electrical signals. The superposition principle applies to the analysis of linear systems that are encountered in electronics, acoustics, and mechanics.

It will help in the explanation of how microphones operate and how electronic circuits function to keep in mind the basic physical interactions that underlie the more complex processes. Our mathematical models of systems are correct only to some level of approximation. In most cases, we can describe a system well enough to understand the basic operations involved in recording sound with simple mathematical descriptions, realizing there are often more subtle factors at play if we carefully scrutinize the processes. A complete picture of the magnetic field generated by a record head and its interaction with tape takes an entire book, but we can still understand how bias current adjustments shape the recovered sound recording by using a quite simplified model. Circuit designers sometimes find they cannot explain what they hear even by detailed analysis, but we are at first concerned with understanding the everyday audio world.

SUGGESTED READING

Halliday, D., Resnick, R., & Walker, J. (2010). *Fundamentals of Physics Extended* (9th ed.). John Wiley & Sons, Inc. ISBN: 978-0-470-46908-8.

Sound

Contents

Acoustics, the study of sound, is a topic that can quickly mystify the casual observer. Fundamentally, it is the physics of how air molecules as a population behave when excited by moving objects. The mathematics used to describe this behavior can be complicated, especially as mathematics isn't our native language. Nonetheless, the behavior itself is conceptually simple enough — as anyone who has tossed pebbles into a pond knows. A rock striking the water's surface causes ripples to spread out in all directions until they bounce off the edges of the pond and reflect back at the same angle from which they arrived, adding and subtracting with each other to produce a complicated pattern on the water's surface. Sound waves behave similarly but with one significant difference: air is far more compressible than water. The springiness of the air allows sound waves to propagate as pressure variations in three dimensions, whereas the water's surface must be deformed to allow the wave to pass. Sound waves are longitudinal waves — that is, the waves of compression and rarefaction move in the direction of wave travel — and water waves, like vibrating strings, are transverse waves in which the movement is perpendicular to the direction of propagation.

THE PHYSICS OF SOUND

Sound propagation depends on the exchange of energy between air molecules. Any gas above absolute zero temperature is in constant motion, with the molecules careening off of each other in a series of random collisions whose total energy is determined by the heat content, which we measure as the temperature of the gas. The higher the temperature, the more agitated the molecules become. Sound energy originates when a vibrating object in contact with the gaseous medium (usually air) forces the air molecules to move with the moving surface. The movement imparts a net velocity to the ricocheting molecules and they bunch up as the moving surface pushes them together in an outward direction. When the molecules move closer together, the air pressure – the cumulative force the molecules exert on each other and any object they encounter – increases. Thus there are two forms of energy contributing to sound propagation: the net velocity of the molecules (kinetic energy) and the pressure they exert (potential energy). Energy can be exchanged back and forth between the two forms, but – as we know from classical physics – it can be neither created nor destroyed. It can be dissipated, though, by expending energy as heat or by doing work, where force is exerted over a distance to move a mass. The total sound intensity at a distance from a sound source is the product of the two forms of energy measured at that point:

$$I = pu = \frac{P}{A}$$ 3-1

where I represents the sound intensity (W/m^2), p is pressure (Pa), u is particle velocity (m/s), P is power (W) and A is the area (m^2) through which the sound energy flows. **Particle velocity** is the rate of movement of the mass of moving air rather than the velocity of the individual molecules. **Sound intensity** is the power of the sound wave measured per unit time flowing through a given area.

As the particles move together, the force they exert on each other increases. If the vibrating surface is oscillating, moving both inward and outward, the force pushing the molecules together begins to decrease, and as the surface reverses direction, moving away from the compressed air region, the particles begin moving backward to fill the region of lowered pressure caused by the retreating surface. Their net movement is therefore zero: they end up back where they started in contrast to the phenomenon of wind, in which the particles move consistently in one direction.

Sound energy travels as successive waves of compression and rarefaction. The **wavelength** (λ) is the distance travelled between successive peaks of compression or rarefaction and is determined by the period of oscillation and the velocity of propagation (not particle velocity):

$$\lambda = cT = \frac{c}{F}$$ 3-2

For a constant velocity of propagation c (m/s), as the frequency f (Hz = 1/s) increases, the wavelength λ (m) decreases. As the propagation velocity increases, so does the wavelength, because the wave then travels farther in the period of one cycle of oscillation. Low-frequency sound wavelengths are quite long (up to 17 m or 56 ft at 20 Hz); high-frequency wavelengths are relatively short (1.7 cm or 0.7 in at 20 kHz), causing sound waves of different frequencies to behave differently in the same acoustical environment. This difference occurs because sound waves interact with objects of similar dimensions in specific ways, as you will see.

When the oscillating air molecules come in contact with other objects, their energy can be transferred to those objects, causing those objects to begin moving. This mechanism is how sound may be detected: the air vibrations cause a microphone or eardrum to begin vibrating in sympathy with the air pressure variations. The sound may then be interpreted as music, speech, or noise depending on the original source of the vibration.

THE PHYSICS OF GASES

To understand more fully the interactions of air with the objects involved in sound generation and perception, consider the physics of energy transfer in a gas. The molecules that make up air are predominantly diatomic nitrogen (N_2) and oxygen (O_2), molecules that rotate and tumble like little dumbbells in space. Their collisions transfer energy between the masses as changes in momentum. Because we are dealing with an immense number of molecular collisions at the scale of interest in acoustics, we treat the molecules statistically as a bulk population of tiny masses that behave as a single, larger mass. We are measuring these cumulative effects of pressure and particle velocity when we convert sound waves to electrical signals.

Air pressure, the force it exerts on a surface, is determined by the number of gas molecules per unit volume and by the temperature. Because gas molecules are in constant motion from the energy associated with the ambient temperature, they bounce around and ricochet off each other continuously, and the relatively large average space between the molecules allows significant compression as well as expansion to take place. The relationship between pressure change and volume change is described by the **bulk modulus** (B) of the gas:

$$B = -\frac{\Delta p}{\Delta V / V}$$
$$3\text{-}3$$

where Δp is the change in pressure and $\Delta V/V$ is the fractional change in volume. The bulk modulus is a measure of the compressibility of the medium and is measured in the same units as pressure, the pascal (Pa = N/m^2 = m·kg/s^2). (The lower the bulk modulus, the more compressible a medium.) The bulk modulus of air is close to 1×10^5 Pa as long as constant temperature is maintained. For comparison, the bulk modulus of water is 2.2×10^9 Pa and that of steel is 1.6×10^{11} Pa, demonstrating their far greater resistance to compression. The bulk modulus and density determine the velocity of propagation v (m/s):

$$v = \sqrt{\frac{B}{\rho}}$$
3-4

where ρ is density (mass/volume = kg/m^3). Because water and steel have much greater bulk moduli than air, sound will travel faster through those media even though their densities are also greater (Figure 3-2). For a constant bulk modulus, however, sound will travel faster through less dense media. Sound will travel faster at higher altitudes, where the air is thinner than at sea level, for example.

The ideal gas law describes the relationship between pressure and volume in a closed physical system:

$$PV = nRT$$
3-5

The gas law states that the product of pressure P (Pa) and volume V (m^3) is constant for a given amount of gas (n in moles [a **mole** is 6.02×10^{23} particles: Avogadro's number, the number of carbon 12 atoms in 12 grams of carbon] and constant absolute thermodynamic temperature (T in degrees Kelvin) [R is the gas constant: 8.3143 J/°K·mol].) In fact, sound pressure waves are adiabatic, meaning the heating generated by compression cannot radiate away in the time between pressure peaks and valleys and therefore temperature does not remain constant. This behavior results because the rate of heat exchange is much slower than the rate of sound propagation. A more precise analysis requires us to consider the change in temperature because there are slight changes in the particle velocity and pressure simply from the localized heating and cooling. An alternative way of considering the gas law as it relates to sound is:

$$PV\gamma = \text{constant}$$
3-6

where γ is the ratio of the specific heat of the gas at constant pressure to the specific heat at constant volume. (The **specific heat** is the amount of heat per unit mass required to raise the temperature by one degree Celsius – a measure of how much energy it takes to heat the gas.) For the predominantly diatomic gases we find in air, $\gamma = 1.4$.

SOUND PROPAGATION

In air, an average pressure is maintained until something perturbs the system. In the case of sound, this can be a compression or rarefaction of the air caused by the movement of an object in contact with the gas. Movement toward us creates a wave of compression; movement away creates a rarefaction. Once such an event occurs, the increase or decrease in pressure radiates through the gas as it moves outward from the source. If the original movement is of a periodic (repetitive) nature, it will set up concentric wave fronts of increased and decreased pressure relative to the average air pressure, and these "shells" will move outward at the speed of sound. Although the individual molecules do not continue to move at that rate – a phenomenon we would call wind (of about 750 miles/hr!) – the pressure variations radiate outward at about 330 m/s [1,100 ft/sec] through the ever-turbulent gas. The individual air molecules do move a small amount as they surge ahead under increased pressure and back in areas of diminished pressure, but there is no net movement. (In fact, there is a very slight net movement of particles because the adiabatic heating and cooling causes the velocity, and therefore the movement caused, to be slightly greater in the compression direction than in the rarefaction direction. This effect is small enough to ignore for our purposes.)

The variations of air pressure we call sounds are minute fluctuations in pressure around a much greater average pressure: the air pressure as measured by a barometer (Figure 3-1). Weather conditions generate much larger changes in air pressure than do sounds. The average air pressure at sea level is around 760 mm Hg, which corresponds to 101 kPa or about 14.7 lb/in². The threshold of hearing is only 2×10^{-8} kPa, so we are talking about exceedingly small variations in air pressure when we consider sound waves. Even the threshold of pain is equivalent to about 2×10^{-2} kPa, so the pressure variations involved with sound are tiny compared to the average air pressure – only a 0.02% deviation for the loudest sound we can tolerate. (The comparatively large changes in air pressure – up to 1 kPa or more – associated with weather don't hurt our ears because for slow changes in air pressure, our ears equilibrate to the

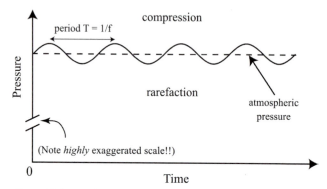

Figure 3-1 Sinusoidal sound pressure wave. The sound waves only vary atmospheric pressure a few hundredths of one percent.

Material	Bulk modulus B (Pa)	Density (kg/m^3)	Speed of transmission (m/s)
Air	1×10^5	1.3	330
Water	2.2×10^9	1×10^3	1.43×10^3
Steel	1.6×10^{11}	7.8×10^3	6.10×10^3

Figure 3-2 The speed of sound depends on both density and bulk modulus.

new average pressure by opening the Eustachian tube to equalize the pressure in the middle ear with the ambient barometric air pressure. Rapid pressure changes can hurt – for example, the changes associated with flying.) Because the density of air decreases with altitude, atmospheric pressure decreases with altitude as well, and the velocity of a sound wave increases in the thinner air.

The shape of the radiating wave fronts depends on the size and shape of the vibrating object and on how far from the source the observer is located. Near a small sound source, the wave is **spherical** (Figure 3–3), and intensity falls off rapidly with distance as it radiates outward through expanding spheres of increasing diameter. The same amount of energy is spread over larger and larger areas as the wave moves away from the source, as the surface of a sphere is $4\pi r^2$ where r is the radius. The result is that sound intensity, measured as power/area (Equation 3-1), decreases proportionally to the inverse of the square of the distance from the source. For this condition to apply, the source must be small with respect to the wavelength of the sounds

Spherical Wave   Plane Wave

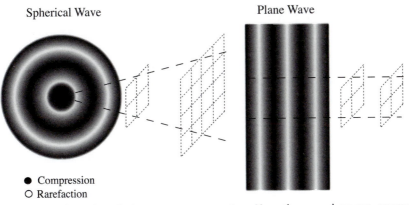

● Compression
○ Rarefaction

Figure 3-3 Spherical and plane wave propagation. Near the sound source, energy flows through an area that increases with the square of the distance from the source. Farther away, energy flows through a nearly constant area for the same change in distance.

involved. Far from the source, the sphere is so large that its surface is nearly a flat plane. For a **plane wave**, intensity decreases linearly as the distance increases because the energy radiates through essentially the same area as the plane progresses.

Another way to analyze the radiation of sound wave energy is to consider the origin of a sound – the initial movement that generates the sound wave. The force that moves the air increases the average velocity of the adjacent molecules. Thus the energy is delivered initially as an increase in particle velocity. The increased velocities cause the molecules to crowd together, raising the local pressure. The sound then propagates away from the source as waves of higher and lower pressure. As the sound energy radiates from the source, it gradually shifts from mostly velocity to a balance of velocity and pressure and then to predominantly pressure. The conditions near a small radiating source where the waves are spherical are known as **near-field** conditions; those farther away are planar and known as **far-field** conditions. As we have seen, these two conditions cause sound waves to radiate differently.

The term "near field" is used to describe loudspeakers designed to be placed near the listener. In this context, "near field" means the listener hears the loudspeaker but not the room reverberations. This setup is also known as **close field**, to avoid ambiguity. In the close field, the loudspeaker may be heard as a point source if there is sufficient distance from the multiple drivers. Any area in which the environment does not interfere with the sound wave is known as **free field**. Such an area exists in part of the close field as well as outdoors or in an anechoic chamber, a room treated to prevent reflections from its surfaces. Farther from the loudspeaker, the reverberant field maintains constant amplitude regardless of distance. The close field occupies the space between the true near field and the far field, allowing the listener to minimize the contribution of the room while hearing the loudspeaker as a single sound radiator.

Equation 3-7 shows the relationship between maximum pressure change Δp_{max} and maximum particle displacement s_{max} in a sound wave:

$$s_{max} = \frac{\Delta p_{max}}{2\pi f c \rho} \qquad\qquad 3\text{-}7$$

We see that the maximum displacement decreases with increasing frequency f, which holds consequences for sensors that depend on displacement to generate output, such as capacitor microphones, as their outputs will also decrease with increasing frequency if not compensated by other factors. The distance traveled by the air molecules during a complete cycle of a sound wave is miniscule. The peak molecular displacement from a sinusoidal sound wave of a painful 130 dB SPL (sound pressure level) is barely 10 μm and at the threshold of hearing on the order of 10^{-11} m. For reference, the diameter of a hydrogen atom is 1.1×10^{-10} m. This tiny displacement will severely limit the mechanical displacement of any sensor, ear, or microphone, so the sensitivity of these systems must be extraordinary.

If we use a **pressure transducer** such as an omnidirectional microphone, only the pressure will interact with the sensing element. If we use a **velocity-sensitive transducer** such as a figure-eight microphone, the particle velocity is converted into electricity as the sensing element moves through the magnetic field to generate the electrical signal. Other types of transducers are sensitive to combinations of pressure and velocity and exhibit spatial sensitivities that reflect the relative balance of the two forms of energy. Directional sensors must be at least somewhat sensitive to particle velocity in order to react differently depending on the angle from which the sound originates.

Though it may not always be obvious, air has significant mass, as evidenced by the damage a windstorm can do. A cubic meter of air has about the same mass as a liter of water. A room of 57 m^3 (2000 ft^3) contains about 68 kg (150 lb) of air. Because the molecules are relatively far apart as they vibrate in space, we do not perceive the mass of the air directly. Nonetheless, there is significant weight to the gas filling the room, and because the particles are in constant random motion they contain significant kinetic energy. The average velocity of the individual molecules can exceed 500 m/s [1100 mph]. The total kinetic energy of the air in the room mentioned previously is an astounding 1.7×10^7 joules; the approximately 1.5×10^{27} molecules contain more total kinetic energy than a 10-ton tour bus traveling at 100 mph! Fortunately, all this turbulent energy is randomly directed and averages out to zero for an observer the size of a human, so we hear only silence until the air is disturbed by a sound source. The random kinetic energy of the air does generate the atmospheric pressure, with which we are in equilibrium.

The mass of air also contributes to its acoustical behavior. In mechanical systems, mass possesses inertia, the tendency to resist changes in motion. In acoustics, the mass of an air column behaves in a similar fashion with the property of **inertance**, whereby a population of air molecules moves together as a single mass. The volume of compressible gas in an enclosure corresponds to a spring in mechanical systems; energy may be stored as changes in air pressure analogous to tension in a stretched or compressed spring. Finally, the gas may act as a resistance through the viscosity of the gas and simply resist the sound wave flow through by friction. It can be helpful to recognize the mechanical equivalents of these acoustical properties in order to analyze the behavior of acoustical systems. These characteristics have electrical analogs as well, and acoustical systems are sometimes modeled using resistors, inductors, and capacitors.

SOUND REFRACTION AND DIFFRACTION

The behavior of sound relative to a room or other confining space is determined by the physics we have discussed. In a room, sound originates from a source and propagates in a straight line until it encounters an object. The waves will then bounce off the surface, bend around the object, be absorbed by the surface, or exhibit a combination of these behaviors depending on the wavelength of the sound and the size, shape, stiffness, and mass of the surface. A stiff surface causes sound waves to **reflect** back at the angle at which they arrived. A compliant surface – one capable of moving in response to the force exerted by the sound wave – can convert some of the sound energy into mechanical motion energy and reflect less energy back into the air. A noncompliant surface will not be moved by the air pressure and will reflect most of the energy back. It is the combination of these effects that determines the "sound" of a room, as some frequencies are reflected and some absorbed, generating the characteristic ambient room sound known as **reverberation**. Ultimately, all the energy is dissipated as heat through molecular collisions and the reverberating sound decays away into silence.

Sounds may bend around physical objects, provided that the wavelength of the sound is longer than the dimensions of the object. Long relative wavelengths bend easily; shorter wavelengths are more beamlike and less able to bend around corners. This behavior can cause sound waves to disperse differently in space depending on the different wavelengths involved. This bending phenomenon is called **diffraction**. Because of their long wavelengths, low-frequency sounds tend to distribute more evenly over long distances than do high-frequency sounds. Figure 3-4 compares diffraction for wavelengths short and long relative to the size of an obstruction.

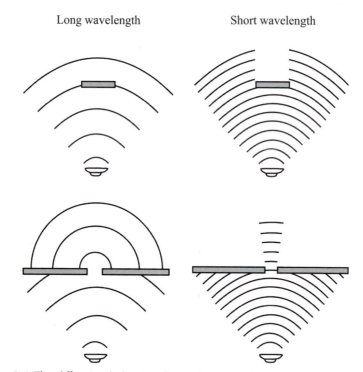

Figure 3-4 The diffraction behavior of sound waves differs at long and short wavelengths relative to the size of an obstruction or an opening.

The flow of sound energy through the air is affected by changes in air density as well as by objects in the path of the sound wave. As the temperature of a gas increases, its density decreases. The speed of sound through air is inversely proportional to the square root of the density of the gas (Equation 3-4), so sound travels somewhat faster through lower-density gas. As sound waves radiate through areas of warmer air, the waves speed up. The result is a bending of the direction of propagation at the interface between lower- and higher-temperature air, a phenomenon known as **refraction**. When the unusual atmospheric condition called **inversion** takes place, higher layers of air are warmer than lower ones. When a sound wave moves upward through an inversion layer, it speeds up and is bent downward. This behavior can make sounds return to earth far from their origin, a potential problem for outdoor music performance venues and noisy factories. Refraction is of minimal concern in the acoustical situations we commonly encounter in sound recording.

ACOUSTICS

Using the physical properties of air, we are able to analyze the acoustics of many of the situations commonly encountered in the process of making sound recordings. One simple acoustic system with broad applications is an enclosure with a hole open to the outside air called the **Helmholtz resonator**. Many musical instruments and some loudspeaker cabinets act as Helmholtz resonators. An empty bottle is a Helmholtz resonator; it oscillates when air is blown across its opening. Resonance occurs when energy is exchanged back and forth between kinetic and potential forms at an optimal rate that maximizes the amplitude of the oscillation. A resonant acoustic system vibrates most easily and produces the maximum vibration at a particular frequency that is determined by the physical characteristics of air and the mechanical characteristics of any objects in contact with the air. If energy is continuously added to a resonant system at the resonant frequency, the energy builds and can even result in catastrophic failure of the structure as it vibrates increasingly vigorously. The energy input reinforces the oscillation when it adds constructively with the existing vibration. For sinusoidal energy, this behavior requires the waves to be in phase and at the same frequency. Figure 3-5 shows how sinusoids add to increase or decrease the output depending on the relative phase of the sinusoids.

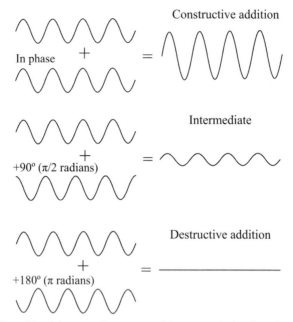

Figure 3-5 Sinusoids of the same frequency add constructively when they are in phase; however, they cancel entirely if they are exactly 180° (π radians) apart in phase. Other phase separations produce intermediate output amplitudes.

By displacing the air in an opening inward, the air in an enclosure is compressed like a spring, increasing its potential energy. When the force compressing the air is released, the mass of air in the opening that was pushed inward tends to move back outward, but it overshoots its original position as the momentum of the moving air continues its movement past its starting point. This movement causes the pressure in the enclosure to drop below its resting value and to begin pulling the air mass back into the enclosure. The combination of the springiness of the air in the cavity, described by the bulk modulus of the gas, and the mass of air in the escape hole or neck together determine the resonant frequency of the system. Equation 3-8 is derived from a physical model assuming harmonic motion that uses the volume and bulk modulus of the enclosed air to estimate the restoring force and the size of the opening to estimate the moving air mass. The derivation assumes that the wavelength of the resonant frequency is long compared to the size of the enclosure so that pressure is equalized throughout the enclosed space.

The resonant frequency of a Helmholtz resonator can be calculated from just the velocity of sound propagation and the dimensions of the enclosure and the opening:

$$f = \frac{c}{2\pi}\sqrt{\frac{A}{V_0 L}} \qquad\qquad 3\text{-}8$$

As shown by Equation 3-8, the Helmholtz resonant frequency (f) is determined by the velocity of sound (c) and by the size of the enclosure and the opening. A is the area of the opening (m^2), V_0 is the static volume of the enclosure (m^3), and L is the length of the opening (m). From Equation 3-8, we see that the resonant frequency increases as the velocity of sound increases. It also increases, although less rapidly, as the area of the opening increases. It decreases as the volume of the enclosure increases and as the length of the neck increases.

ROOM MODES OF REFLECTION

When a sound wave at the resonant frequency excites a resonant object, it vibrates easily with the incoming wave; at other frequencies, it takes more energy to cause vibration. This phenomenon applies equally to large spaces like rooms and to small enclosures like musical instruments and even to ear canals. Understanding the resonant acoustical behavior of these spaces will help us explain many aspects of how we perceive the sounds around us and how best to record them.

When we hear sounds in a room, we hear the resonant behavior of the room structure superimposed on the original sound. The room acts as a filter as it alters the balance of frequencies that make up the exciting wave, reflecting some and absorbing others. Within a room, the wavelength of a sound wave will determine exactly how a sound and its reflections interact. When the wavelength (or a multiple or fraction of it) corresponds exactly to a room dimension, the sound wave will reinforce and cancel its reflections in a pattern of stationary pressure peaks and troughs. Most rooms are not designed for desirable acoustical properties; they are designed to serve the general needs of the occupants with little consideration of acoustics and often the dimensions and shapes that prove most useful for general human activities are particularly bad for music reproduction.

Because rooms are generally rectangular, they have three main axes: front-to-back, side-to-side, and floor-to-ceiling. The lengths of the axes determine the wavelength of a sound wave that will reinforce constructively (see Figure 3-5) as it bounces from one surface to the other. Each distance traversed by a sound wave determines a resonant **room mode**, or mode of oscillation (Figure 3-6). The sound pathways along the three main axes produce modes known as **axial modes** in which a single bounce completes the transit. Paths that incorporate four bounces among these reflecting surfaces are known as **tangential modes**, and paths that generate six

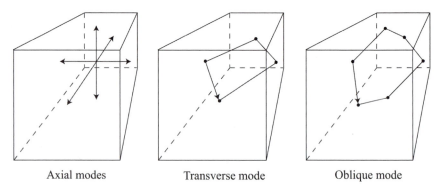

| Axial modes | Transverse mode | Oblique mode |

Figure 3-6 Room modes showing possible paths taken by sound waves reflecting from room surfaces. Simple modes include fewer bounces than more complex modes and consist of shorter path lengths.

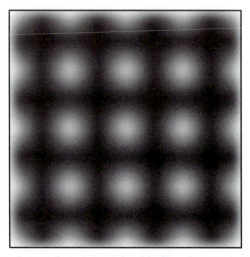

Figure 3-7 Overhead view of pressure nodes (black) and antinodes (white) in a square room. This standing wave is mode 4,4,0.

bounces are called **oblique modes**. Each mode results in sinusoidal interferences at frequencies determined by the ratio of the path length to the sound wavelength, creating the characteristic resonant behavior we hear. Not only do paths at exactly the sound wavelength reinforce each other; so do multiples of the wavelength, resulting in a complicated series of resonant modes. The mode frequencies may be calculated from the room dimensions in Equation 3-9:

$$f_{n_x, n_y, n_z} = \frac{c}{2} \sqrt{\frac{n_x^2}{L^2} + \frac{n_y^2}{W^2} + \frac{n_z^2}{H^2}} \qquad\qquad 3\text{-}9$$

where c is the speed of sound; L, W, and H are the length, width, and height of the room; and $n_{x,y,z}$ are integer multiples 0, 1, 2…n. Even simple rooms shapes have complicated modes.

As sound waves in a room reflect from walls, floor, and ceiling, regions of increased pressure will coincide in certain zones, as will regions of pressure cancelation. For a constant sinusoidal sound wave, this behavior results in specific areas of greater sound pressure and areas of diminished sound pressure distributed regularly throughout the room. Areas of cancellation, where sound pressure is lowest, are called **nodes**; locations at which sound pressures combine are called **antinodes**. As you walk around the room, you can hear these areas clearly; where you stand influences what you hear. This phenomenon of reinforcement and cancelation of sound pressures is known as **standing waves** because the areas remain stationary when excited by a continuous sinusoidal sound wave.

Reflected sound waves tend to reinforce incident waves near boundaries such as walls. In corners, where multiple surfaces reflect, energy builds, especially at long wavelengths. Placing loudspeakers near these boundaries increases the low-frequency energy projected back into the room in addition to creating standing waves. Special absorbers called **bass traps** can be used to dissipate some of the excess low-frequency energy. Standing wave intensity may also be reduced by constructing rooms where the dimensions are not multiples of each other, reducing the amount of mode reinforcement.

Of course, most real sounds are not simply sinusoidal but are rather composed of many different sinusoidal components of varying amplitude and phase. When musical sounds combine in a room, the resulting ever-changing patterns of standing waves can be extremely complex. We are used to listening to music performed and often recorded in such an environment.

Room acoustics incorporated in reproduced sounds cause them to sound natural to us. Recognizing that sound is not uniformly distributed in most rooms is a step toward understanding why the sound produced by a microphone can be so sensitive to tiny changes in position.

SUGGESTED READING

Olson, H. F. (1967). *Music, Physics and Engineering* (2nd ed.). Dover. ISBN: 0-486-21769-8.
    A great resource for musical acoustics, despite its age.

Baranek, L. L. (1996). *Acoustics*. Acoustical Society of America. ISBN: 0-88318-494-X.
    A definitive acoustics reference that uses the electrical model analysis of acoustical
    systems.

Everest, F. A., & Pohlmann, K. C. (2009). *Master Handbook of Acoustics* (5th ed.).
    McGraw-Hill. ISBN: 978-0-07-160332-4. Chapters 9–13 discuss sound and its
    interaction with the surrounding environment.

CHAPTER FOUR

Hearing

Contents

The study of the nervous system's cognitive response to sound stimuli is known as **psychoacoustics**: it is part acoustics and part psychology. The visual system is often considered the more important sensory modality, but the auditory system is far faster in its analysis and response to incoming sensory information. It is when we first begin to work with sound recording that we become aware of many of the subtleties present in our auditory system. For example, phenomena such as masking, in which only the louder of two sounds close together in frequency is perceived, are attributable to the behavior of our auditory physiology. All one needs to do to appreciate our native sound-processing capabilities is listen to how different the world sounds through microphones and headphones rather than just through our ears. A complete study of the function of the auditory system is exceedingly complex and beyond the scope of this discussion; however, we can still appreciate some of the features of the system that affect directly how we perceive sounds, especially when they are critical to the processes employed in the recording of sound.

Our auditory system is incredibly sensitive, allowing perception over many orders of magnitude in both amplitude and frequency. We can discriminate tiny changes in sound timbres and accurately determine where in space a sound originates. We experience sound through our ears and nervous system, so our perception cannot be divorced from these mechanisms, and their characteristics influence what we hear. The pattern of air pressure vibrations striking our ears provides the input, but just as the structure of a room alters sound waves as they pass through it, so the apparatus of hearing changes the information we ultimately receive in our brains' auditory processing areas. We have the advantage of an adaptive brain, which learns how to process the sensory inputs we receive from the

The Science of Sound Recording
ISBN 978-0-240-82154-2

43

cochlea, the organ that converts sound wave energy into neuronal signals, and adapts to the inherent imperfections of our own auditory system. Nevertheless, we must process sound information through our hearing organs, and our perception includes distortions – alterations in the way sounds are transmitted and converted to neuronal signals and in the way our brain interprets these inputs and renders for us what we refer to as hearing.

THE AUDITORY SYSTEM

Musical sounds contain a complex combination of individual sinusoidal components, each with a particular amplitude, frequency, and phase relationship with the other elements. By combining these components, the characteristic sounds of different instruments and voices are created. These combinations are known as **timbre**, a quality that distinguishes the sound of different instruments even when they play the same note at the same loudness. This explains in part why a horn sounds different from a stringed instrument: each produces a specific combination of mathematically related frequencies known as **harmonic overtones** that result in the characteristic sound. (Some sounds, such as bells, contain non–harmonically related overtones.) The timbre of a sound must be preserved in the recording process in order for a sound to be perceived as natural sounding. Timbre alone is not sufficient to fully discriminate between instrument sounds, however, as the time course of onset and decay of notes is also characteristic of instrument sounds. Instruments with similar timbres will sound different from each other if their attack, sustain, and release characteristics are different.

We often assume that what we perceive as pitch is exactly equivalent to the actual vibratory frequency of the sound wave and that what we perceive as loudness is directly proportional to the amplitude of the sound wave pressure variations. In fact, the operation of our auditory system deviates somewhat from these ideals, and we must factor these deviations into our understanding of the process of hearing: the first stage in the process of hearing, handled by the outer ear (**pinna**) and **ear canal** (Figure 4-1), distort the incoming pressure wave intentionally. The ridges of the ear reflect particular frequencies from certain directions in order to create an interference pattern that can be used to extract information about the elevation from which a sound originates. Sounds originating from above are reflected with increased high-frequency content relative to the same sound

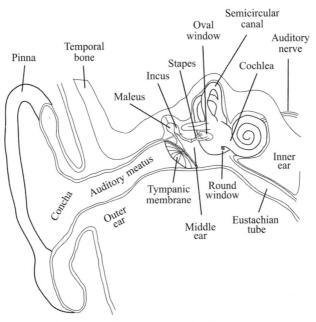

Figure 4-1 The anatomy of the human ear.

originating at the level of the ear. Front-to-rear discrimination also depends in part on the shadowing of rear-originating sounds by the pinna. The **external auditory meatus**, or auditory canal – the guide conducting the sound wave to the eardrum – is a resonant tube that further alters the frequency balance of the sound wave. The resonant frequency falls in the same frequency range as the peak in our sensitivity, around 3kHz, and creates a maximum boost of about 10 dB. This is the frequency range that conveys much of the information contained in speech; in fact, the wired telephone system transmits frequencies from only 300 to 3400 Hz. The vibrations that finally excite the eardrum differ from the original sound pressure variations: they contain an altered balance of frequency components, affecting the timbre of the sound.

The **tympanic membrane**, or eardrum, is a flattened conical membrane that stretches across the inner end of the ear canal. It is open to the auditory canal on the outside and in contact with a set of three tiny middle ear bones, the **ossicles**, on the inside. The pressure on the outside of the tympanic membrane is determined by the sound wave and the static atmospheric pressure. On the inside of the tympanic membrane, the static air pressure in the middle ear is equilibrated through the **Eustachian tube** to the throat, while the sound vibrations are conducted through the bones to the cochlea. Equilibrating middle ear pressure with the outer ear pressure reduces any damping of the tympanic membrane caused by pressure differences that result in unequal forces on opposite sides of the membrane. The mechanical characteristics of the middle ear allow for active control of the transmission efficiency between the eardrum and the cochlea. The tiny muscles connecting and suspending the bones and ligaments can contract, stiffening the connection and drawing the ossicles away from their attachments to the tympanic membrane and cochlear oval window. This contraction allows adjustments of the sensitivity of the hearing process through the **acoustic reflex**, which may be activated by loud sounds (to protect the inner ear from possible damage) as well as by the intention to begin vocalizing. The acoustic reflex mechanically reduces the dynamic range of the input to the cochlea, much like a compressor or limiter. The time course of activation and release of the reflex can be used to determine the attack and release of electronic compression characteristics so that they sound natural.

The primary function of the bones of the **middle ear** is to amplify mechanically the airborne vibrations in preparation for transfer to a liquid medium. Because liquids are denser than gases and less dense than solids like bone, we encounter a potential problem when converting the energy in one medium to energy in another: the systems require different amounts of force to drive them. The bones act to focus the vibrations of the relatively large eardrum and deliver them efficiently to the small oval window of the cochlea as well as to protect the cochlea from too much input. They act as an impedance converter, efficiently coupling the low-impedance air pressures with higher-impedance liquid pressures inside the cochlea.

Because the cochlea is stimulated by mechanical vibrations, it may be activated by vibrations of the surrounding temporal bone that do not come through the ear canal, an effect known as **bone conduction**. Although the strength of the bone conduction is well below that of sounds conducted through the middle ear bones, it is still audible. It partially explains why our voices sound different in recordings from when we are vocalizing – recordings do not contain the internally conducted sound we hear through bone conduction.

The cochlea (Figure 4-2) is a dual-purpose structure: it converts mechanical vibrations into neuronal electrical signals and separates the frequency content of the incoming sound into discrete frequency bands. It functions like a spectrum analyzer, a device that breaks sounds or other

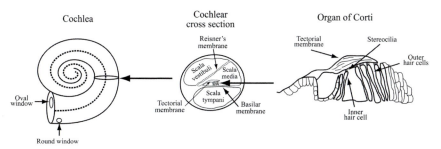

Figure 4-2 The structure of the cochlea. The oval window is in contact with the scala vestibuli and the round window is at the far end of the scala tympani. The scala vestibuli and scala tympani are connected at the apex of the cochlear coil.

signals into their discrete frequency components. It is, however, an imperfect analyzer, and there is some overlap between the bands whereby strong signals in one band slightly stimulate adjacent ones, creating harmonic distortion. It is up to the brain to sort out the raw data from the cochlea and produce the sensation we call hearing.

COCHLEAR PHYSIOLOGY

The cochlea performs the task of separating different frequencies by providing an array of sensing cells, the **inner hair cells**, that are mechanically stimulated by the movement of the two membranes between which they are connected. The membranes are caused to vibrate by the fluid filling the chambers, which is in turn caused to vibrate by the bones of the middle ear. Different areas of the cochlear membranes displace maximally at different frequencies as the traveling wave from the oval window moves toward the apex, separating the areas maximally stimulated by different frequencies of vibration along the cochlea's length. High frequencies stimulate the area nearest the oval window, and low frequencies drive the far end. Due to its placement, each inner hair cell responds to vibrations of a specific range of frequencies, and each is connected to a nerve that sends signals to the brain. The bases of the hair cells are attached to the **basilar membrane**, and projections from their apical ends called **stereocilia** attach to a second overlying membrane, the **tectorial membrane**. As the cochlear fluid vibrates, it produces relative movement of the membranes, causing the stereocilia to shear, which opens ion channels in the hair cells that then depolarize the affected cells.

The voltage across cell membranes is determined by the distribution of positive and negative ions on each side of the membrane, usually with more sodium and calcium on the outside and more potassium on the inside. In the resting state, these ion concentrations are maintained by cellular metabolic activity so that cellular interiors are negatively charged relative to the extracellular fluids. In free solution, ions distribute themselves to maintain a constant and evenly dispersed concentration. When membranes impermeable to the ions separate the inside and outside concentrations, a voltage is created by the concentration gradient across the membrane with a separate contribution from each specific ionic gradient. **Ion channels** embedded in the membrane that allow specific ions to pass through the membrane may be opened by chemicals, the voltage across the membrane, and, in the case of inner hair cells, by mechanical means. When ions are allowed to flow down their concentration gradient through the membrane, the membrane voltage decreases and the cell is said to **depolarize**. When the membrane potential depolarizes enough to reach a threshold voltage, the hair cell releases excitatory neurotransmitter chemicals that trigger an **action potential** in the adjacent nerve fiber that is then conducted to the brain through the auditory nerve.

The cochlea is known to be extremely selective in its tuning, a feature thought to be accentuated by an active feedback mechanism. A second variety of hair cell, the **outer hair cell**, is also attached between the basilar and tectorial membranes, but these cells receive neuronal inputs from the brain. They are thought to affect the tension of the tectorial membrane in localized areas, effectively increasing the tuning sharpness of the cochlea. In fact, sounds actually originate from the cochlea and may be heard in the ear canal, a phenomenon known as **otoacoustical emission**. It appears that this system may be partly responsible for a form of distortion inherent in the auditory system, the generation of sum and difference frequencies that occur when two separate sinusoidal sounds are presented to the ear.

Prolonged stimulation of the cochlea can deplete the metabolic energy available to the hair cells, an action that can shift the threshold of hearing based on the loudness of incoming signals. This process can have a long recovery time, producing a phenomenon called **threshold shift**. After exposure to loud sounds, the sensitivity of the auditory system can be reduced for hours, a fact often observed after hours of playing or mixing loud music. The delicate hair cells are susceptible to physical damage from excessive excitation. Overexposure to loud sound can destroy these cells, leading to permanently diminished hearing sensitivity, as the individual frequency detectors are damaged. Although early research indicates that some regeneration might be possible, the condition should be considered irreversible and avoided at all cost. Reducing exposure to loud sound should be a primary consideration for everyone working in audio.

PERCEPTION OF SOUND

Exactly how our brains process auditory inputs and create for us the conscious awareness of hearing sound is still within the realm of mystery. Although research has elucidated much of the structure of the neuronal pathways and processing centers in the brain, the explanation of exactly how we perceive sound is still incomplete. Fortunately, for the understanding required to become adept at sound recording, we need only appreciate the operational characteristics of the system as it applies to how we perceive sound.

There are critical features of our auditory system that we must consider to appreciate which characteristics of sounds are necessary to preserve in our recordings. In order to preserve the cues we use in localizing the positions of sound sources in space, we need to understand how we tell where in space sounds originate. To make accurate sound recordings, we need to be aware of how we perceive the amplitude and frequency information in sound waves. To convincingly manipulate sounds in the studio, we need to know how sounds behave in our environment.

Any sound originating in space reaches us through our two ears. Because they are separated by several inches, sounds not originating directly ahead or behind reach them at slightly different times. Further, they strike the closer ear with slightly more energy, making that side sound louder. By using these relative time-of-arrival and loudness cues, we determine where in space a sound originated. We can make use of these cues to fool the ear, as we might when mixing sounds in the studio, placing sounds in different apparent positions in a mix. Preserving these cues is critical to making stereo recordings that capture the realism we desire. Sound striking one ear first is a very strong cue to the position of a sound source. By delaying an element panned to one side in mix by a few milliseconds, it can take many decibels of gain to make that sound seem as loud as the same element panned undelayed to the opposite side. The pan controls built into most mixers use only the apparent loudness cue to pan signals from left to right, ignoring the time-of-arrival difference.

The phenomenon of masking is also important in sound recording. Cochlear regions that respond to the same frequency range are called **critical bands**. These bands are not fixed filters but rather localized areas of the cochlea that vibrate maximally at a characteristic range of frequencies. The width of critical bands increases with increasing frequency; resolution is sharper for low frequencies. Each band contains over a thousand individual hair cells that respond when that area of the tectorial and basilar membranes are deformed by vibration. Each critical band responds to the sum of the energy in that frequency range, so we perceive only the louder of two sounds of similar frequency as long as there is a sufficient level difference between the sounds. This principle allows some types of noise reduction processing and methods of audio data compression like mp3s to work. It also complicates mixing two sounds with similar frequency content. With the increasing reliance on masking curves, a frequency scale called the **Bark scale** was devised to reflect the cochlear critical bands. The Bark scale has 24 bands that correspond to cochlear critical bands. Figure 4-3 shows the correspondence between frequency and the Bark scale. The curve reflects the widening of critical bands as frequency increases. The Bark scale is used in perceptual audio coding research.

Our perception of timbre depends on the frequency response behavior of our auditory system. Because we do not perceive all frequencies to be equally loud at the same intensity, discussing loudness demands a measure of loudness that takes the sensitivity to frequency into account. A unit known as the **phon** is used in loudness comparisons; it is the intensity (sound pressure level in decibels) of a 1kHz sine wave tone judged to have the same loudness as the signal in question. A similar unit, the **mel**, relates absolute

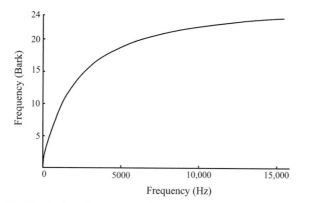

Figure 4-3 The Bark scale translates frequency into cochlear critical bands.

frequency to perceived pitch. Both perceived pitch and loudness deviate from linearity with regard to frequency and amplitude.

Due to the physical characteristics of the auditory system, we do not perceive all frequencies as equally loud for the same sound pressure level. Further, the variation in perceived loudness as a function of frequency changes with the loudness level. The **curves of equal loudness**, often called Fletcher–Munson curves after the original researchers, show the effect of sound level on the spectral sensitivity of our auditory system. The measurements vary significantly among individuals and are average values. Fletcher and Munson used headphones and pure tones in their work; Robinson and Dadson later used loudspeakers to reproduce the pure tones in an anechoic room. The two sets of curves differ, but both show a peak of sensitivity at around 4 kHz, near the resonant frequency of the auditory canal, and a significant decrease in sensitivity at extreme frequencies, especially low frequencies. Figure 4-4 shows the latest data obtained by the ISO (International Organization for Standardization), which differ somewhat from the earlier research, especially in the lower frequencies. Using pure tones to determine sensitivity will produce different curves than those produced using noise bands that stimulate the cochlea more broadly, because

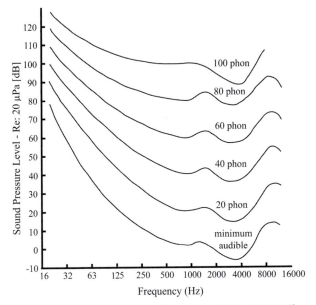

Figure 4-4 Revised equal loudness curves (data from ISO 226:2003). The curves show how much sound pressure is required to sound equally loud at each frequency.

each hair cell responds to a range of frequencies (its critical band), more of which is stimulated by narrow-band noise than by a pure tone. Also, as frequency increases, more energy is contained in each critical band, somewhat increasing the sensitivity with increasing frequency relative to measurements using pure tones.

The flattening of the sensitivity curves has important implications for sound mixing: the level at which we listen affects our perception of the balance of frequencies present in the signal. As can be seen from Figure 4-4, the equal loudness curves flatten out at around 80 dB SPL, and not coincidentally, that is close to the loudness often designated as the standard listening level, 85 dB SPL. By mixing at that average loudness, the spectral balance will be perceived to be correct at usual listening levels. When listening at low levels, low frequencies are particularly inefficient, and there is thus a tendency to boost them. When that mix is played at higher listening levels, the bass will sound accentuated. It is therefore important to listen and mix at levels close to the intended listening level.

REFERENCES

Acoustics—Normal equal-loudness-level contours. (2003). International Organization for Standardization. ISO 226.

Kessel, R. G., & Kardon, R. H. (1979). *Tissues and Organs: A Text-Atlas of Scanning Electron Microscopy.* W. H. Freeman and Company. ISBN 0-7167-0091-3.

Robinson, D. W., & Dadson, R. S. (1956). A re-determination of the equal loudness relations for pure tones. *British Journal of Applied Physics, 7,* 166–181.

SUGGESTED READING

Gulick, W. L., Gescheider, G. A., & Frisina, R. D. (1989). *Hearing: Physiological Acoustics, Neural Coding, and Psychoacoustics.* Oxford University Press. ISBN 0-19-504307-3.

Pickles, J. O. (1988). *An Introduction to the Physiology of Hearing.* Academic Press Limited. ISBN 0-12-554754-4. Chapters 1–5 detail the physiology; later chapters address the central nervous system and psychoacoustic principles based in the physiology of the auditory system.

CHAPTER FIVE

Electronics

Contents

In order to manipulate and store sounds, we need a representation that can be easily transferred from point to point, altered to suit our wishes, and stored in a permanent fashion. A simple and flexible such representation is the flow of electrons, or electricity. We use an electromechanical device, a microphone, to convert the air pressure variations of sound into an analogous flow of electrons. It is then simple to distribute the electrons through a wire to other places. It is also possible to convert the flow of electrons into magnetic flux that can be stored for a long time when retained as magnetic patterns on media such as analog tape and digital hard disks. Electronic circuits are involved in the majority of sound recording operations.

Devices including recorders, amplifiers, mixers, equalizers, dynamic range processors, and delay-based effects processors give the engineer great ability to manipulate artistically the acousmatic reproduction of recorded sounds. Although these devices are complex, their operation is based on a few elementary concepts of analog electronics. **Analog electronics** refers to circuits in which the signal varies continuously, in contrast to digital systems that sample the signal at regular intervals. Until recently, analog was the only method available for processing audio data. Much modern sound recording and mixing involves computers, but analog electronic devices are still widely used in sound recording, especially as microphones, loudspeakers, preamplifiers, and mixers. Understanding electronic circuits starts with electricity.

The Science of Sound Recording
ISBN 978-0-240-82154-2

53

BASIC ELECTRICITY

Electricity is simply the movement of **charge** (symbol q). Charge is a fundamental property of matter that has two possible states: positive and negative. (Neutral or zero charge may also be considered a state, though this is not of interest in electronics.) Like charges repel each other; opposite charges attract and can combine and cancel. Charge was theorized before the structure of the atom was understood, and it was initially assumed that positive charge flowed through electric circuits. Electric current is now known to be due predominantly to the flow of negatively charged electrons; however, current can be conducted by positive charges in semiconductors and in ionic solutions. The flow of charge can be through a resistive medium like air (lightning), through a solid conductor like a metal wire, or through a semiconducting material like silicon. The path through which the charge moves is called a **circuit**. The charge moves from a source through a loop of circuit elements and back to the source. For **direct current** (DC), the charge flows in only one direction (although its magnitude may continuously vary); for **alternating current** (AC), the charge flows back and forth in both directions (Figure 5-1).

Charge (q): Charge is the carrier of electric current. Electric current usually consists of loosely bound electrons of the conductor that are relatively free to move. Charge can also be static, which occurs if an excess of one charge, positive or negative, accumulates on an object. Charge is measured in **coulombs**, a quantity of electric charge equivalent to 6.414×10^{18} electrons, the amount of charge transferred by one ampere in one second.

Current (i): Current is the flow of charge. Current is $i = dq/dt$, the amount of charge flowing Δq per unit time Δt, the rate of charge transfer. Current is measured in **amperes** (A) with the units of coulombs/sec.

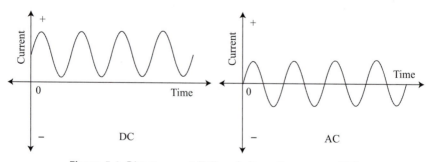

Figure 5-1 Direct current (DC) and alternating current (AC).

Voltage (v): Voltage is the force propelling charge through a circuit, like pressure in a fluid system. Voltage is defined as $v = -W_\infty/q$, where W_∞ is the work that would be expended to move a charge from infinity to the point of its measurement. Voltage is also called electromotive force (EMF) as an indication of its nature as the driving force of electricity. Because of the amount of work potential a voltage possesses, it is also often referred to as **potential.** Voltages measured in a circuit are differences in potential between two points, so we are really talking mostly about potential differences when we refer to voltages. Audio signals are most often represented as time-varying voltages. Volts have the units of joule/coulomb.

Ground: Because voltages are differences in electromotive force between two points, we need a standard reference potential against which to measure the potential at each point in a circuit. The reference used is ground, ultimately derived from the earth, in theory at least. Because the earth is extremely large (~6×10^{24} kg) relative to electric circuits, it is able to absorb or provide an essentially infinite amount of charge without changing its potential. The ground reference in a circuit is connected through power line wiring to the earth at a point where power lines enter the building. This reference voltage is taken to be zero volts, and circuit voltages are measured against that reference. The ground connection also plays an important role in conducting unwanted induced currents away from our audio circuits. Current will flow through the lowest resistance path to ground, whether through an intended circuit path or an accidental one.

Impedance (Z): Impedance is an opposition to the flow of current (analogous in effect to the diameter of a pipe in a fluid flow system) and is measured in ohms for resistance (R) and **reactance (X)**, two types of impedance. Resistance does not depend on the frequency of the signal; reactance does. The equivalent resistance of a reactance is a function of frequency; however, for a given frequency, the magnitude of a reactance can be considered much like a resistance when analyzing a circuit. At the same time, any reactive impedance also has a frequency dependent effect on the propagation time through the circuit and can alter the relative phases of sine wave signal components of different frequencies. This behavior complicates the analysis of reactive circuits with time-varying signals.

When we wish to transfer energy between two systems, their impedances must be equal for optimal transfer. When the impedances are mismatched, energy is reflected back from the junction instead of flowing through.

Impedance matching is relevant to mechanical and acoustical systems as well as electrical ones.

Power (P): Power is the amount of work done (or energy expended) per unit time in a circuit and is given by:

$$P = i \times v = \frac{v^2}{R} = i^2 \times R \qquad\qquad 5\text{-}1$$

where Ohm's law (see Equation 5-8 later in this chapter) has been substituted for either i or v.

Electric and magnetic fields ($\vec{E}$ and $\vec{B}$): Every charged particle creates an electric field ($\vec{E}$) that diminishes in magnitude as it radiates outward in all directions. Any two separated unlike charges, called a **dipole**, generate between them an electric field whose strength and direction depend on the spatial location of the measurement relative to the charges. This field is described mathematically as a **vector field**, a large collection of individual vectors, each of which represents the electric force magnitude and direction at that particular point in space. The field exerts a force on any charged particle within it. Lines of equal force radiate out in every direction from the charges, and the longer the path, the lower the strength of the field along that path.

Although we normally regard a wire simply as a conduit for electrical current flow, there is a very important phenomenon generated by current flowing in a wire, namely, the creation of a magnetic field ($\vec{B}$) that varies with the changing flow of current. For a current flowing in a hypothetical infinite-length straight wire, the magnetic field magnitude B at a distance R from the wire is given by:

$$B = \frac{\mu_0}{2\pi R} \qquad\qquad 5\text{-}2$$

where μ_0 is the permeability constant (1.26×10^{-6} T-m/A) and i is current (A). As the current increases, so does the field strength; as the distance increases, the field strength decreases. Whenever current flows, it sets up a magnetic field, and any time a wire moves through a magnetic field, a current flow is produced in the wire. Known as **induction**, this phenomenon allows circuits to be coupled with no physical connection by a transformer, where two coils share an overlapping magnetic field. The inherent connection between electric current and magnetic fields is at the heart of two critical stages of sound recording: the conversion of sound into electric current in microphones (and the converse in loudspeakers) and the magnetic recording process used in both analog and digital recorders. It can also create problems by introducing contamination from power lines and electrical devices.

Signal: The term "signal" denotes a time-varying voltage or current that encodes information, often a voltage or current that varies in proportion to a measured quantity, such as air pressure.

PASSIVE ELECTRONIC DEVICES

The simplest **electronic components** are the passive devices: resistors, capacitors, and inductors. "Passive" means that they do not require external power to function – only the power contained in a signal itself. (Active components such as transistors, vacuum tubes, and integrated circuits require external power that can be modulated by the signal.) Electronic devices are characterized by their current-voltage (i-v) relationships; as we vary the current through the device, what does the voltage across them do? These circuit elements all exhibit some deviations from the "ideal" behavior, but these differences are small enough to ignore for now.

Resistors

$$v = iR \qquad\qquad 5\text{-}3$$

Resistors are passive devices that have constant impedance regardless of the frequency of the current flowing. They oppose the flow of current by an amount directly related to the resistance (like a constriction in a hose resists the flow of liquid). Resistances are the simplest of the passive devices, and working with purely resistive circuits is a good place to start to build an understanding of electronic circuits. Resistances in series simply add arithmetically:

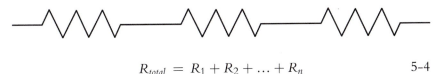

$$R_{total} = R_1 + R_2 + \ldots + R_n \qquad\qquad 5\text{-}4$$

while parallel resistances add as:

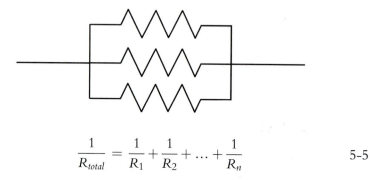

$$\frac{1}{R_{total}} = \frac{1}{R_1} + \frac{1}{R_2} + \dots + \frac{1}{R_n}$$ 5-5

Using just resistors, circuits are able to attenuate signals only linearly, but the variable resistor, or potentiometer, is a purely resistive device at the center of most audio processes: it is the volume control or fader. The **potentiometer** ("pot") consists of a fixed resistance from end to end with a movable wiper that can connect anywhere between the ends of the resistive element (see Figure 5-2).

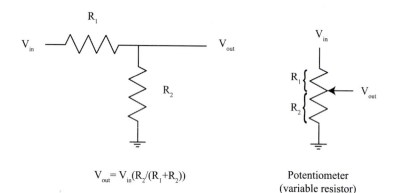

$$V_{out} = V_{in}(R_2/(R_1+R_2))$$ Potentiometer
(variable resistor)

Figure 5-2 The voltage divider and potentiometer.

Capacitors

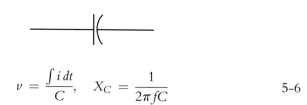

$$v = \frac{\int i\,dt}{C}, \quad X_C = \frac{1}{2\pi fC}$$ 5-6

A **capacitor** consists of two electrically charged conductive surfaces called **plates** placed close together with a dielectric (nonconducting) material between them. When a charge is applied to one plate, it repels like charges on the opposite plate, leaving a net opposite charge. The voltage that builds across a capacitor is related to the charge that resides on the capacitor's plates by the equation $q = Cv$, so the voltage is equal to the charge divided by the capacitance ($v = q/C$). The total charge q is the sum of current flow or $\int i\, dt$. Because current is the rate of charge flow, the larger the current, the faster the voltage builds. For direct current, the capacitor charges with a time constant that depends on the capacitance value and the impedance through which the current flows into the capacitor, because that impedance will determine the maximum amount of current flowing into the capacitor from a given voltage input. Once the capacitor is fully charged, no further current flows. The capacitor is thus an effective block for direct current. For alternating current (such as audio signals), the response is more complicated: the voltage that develops across the capacitor depends on how quickly the current is changing in magnitude and direction. Because it takes time for the charge to build up, there is a frequency dependent delay (or phase shift) in the voltage across the capacitor. Capacitive reactance X_C is inversely proportional to frequency, as shown in Equation 5-6.

The unit of capacitance is the **farad** (F). One farad equals one coulomb/volt. Until the advent of monster car stereo systems, capacitors of this size in audio devices were unusual; values of microfarads (mF), nanofarads (nF), and picofarads (pF) are most common. A charged capacitor acts like a battery, though it can generate a current only until it is fully discharged.

Inductors

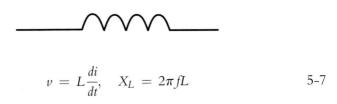

$$v = L\frac{di}{dt}, \quad X_L = 2\pi fL \qquad\qquad 5\text{-}7$$

An **inductor** is most often a simple coil of wire that can be wrapped around either an air or metal core. As current flows into an inductor, a magnetic field is created around the coil. When the current stops, the magnetic field collapses, generating an induced current flow in the coil in the opposite direction to the original current. Low-frequency currents flow easily through the inductor, but as the alternating current frequency increases, the impedance of the inductor increases. Like the capacitor, the inductor introduces a phase shift. From Equation 5-7, it is evident that inductive reactance X_L increases with increasing frequency – the opposite of the results with a capacitor.

The unit of inductance is the **henry** (H). A current changing at a rate of one ampere per second through one henry of inductance produces one volt across the inductor. Most electronic circuits employ much smaller inductances than one henry.

Transformers are special types of inductors in which two separate coils share overlapping magnetic fields. When the primary coil is driven, it generates a magnetic field that induces current to flow in the secondary coil. Because there is no physical connection between the primary and secondary coil wires, the two circuits are physically isolated from each other. Often, transformers consist of an iron core wrapped with two or more coils that couple through the magnetically susceptible metal. Transformers are used to get voltage gain (at the expense of current reduction) and to step down power line voltages for power supplies. Transformers are also used to match impedances between devices and to provide ground isolation.

Ohm's Law

Ohm's law is one of the simplest yet most important principles of electronics:

$$v = iR \qquad\qquad 5\text{-}8$$

The voltage drop across a resistor is the current through it multiplied by its resistance. This equation holds true for the impedance of inductors and capacitors as well, if we take into account their frequency-dependent nature.

Kirchhoff's Laws

In addition to Ohm's law, the work of Gustav Kirchhoff in the 1840s details a couple of important behaviors inherent in electronic circuits. Kirchhoff states that the amount of current entering a circuit junction exactly equals the amount of current leaving that junction. In other words, current is conserved. A second rule states that the sum of voltage drops around a complete circuit is equal to zero. For example, a battery connected to a circuit produces a fixed voltage and the sum of the voltage drops around the circuit connected to the battery exactly equals that same voltage but in opposite polarity. These two rules make the analysis of circuits possible.

Although they may seem obvious to us, they are important to understand as principles of electric circuits.

Although most electronic devices are full of active components like op-amps (explained later in this chapter) and transistors, much of the actual circuit function is accomplished by simple arrangements of the passive elements: resistors, capacitors, and inductors. Understanding of these simple circuits allows an audio device user to examine the schematic diagram and quickly gain knowledge about the function of the device. It also makes troubleshooting possible.

The simplest functional circuit, but one of the most important, is the **voltage divider** (Figure 5-2). The voltage divider is so called because the input voltage divides in proportion to the resistances of the circuit. If we measure the voltage drop across R_2, it is the input voltage times the ratio of R_2 to the total resistance of the circuit (ignoring the effects of other devices connected to the circuit). A variable resistor is in fact an adjustable voltage divider that lets us control levels by continuously varying the ratio between R_1 and R_2.

The voltage divider helps explain a general principle relating to the proper interconnection of devices: impedance matching. When we connect two audio devices, we wish for the signal voltage at the output of the first device to be received at the input of the second device with as large amplitude as possible. In order to guarantee this result, the input impedance of the second device must be greater than the output impedance of the first device. These impedances can be modeled with R_1 representing the output impedance and R_2 representing the input impedance. We can see that when R_2 is much larger than R_1, most of the voltage will drop across R_2 as we desire; in order to efficiently transfer voltage, the input (load) impedance must be large relative to the output (source) impedance. Figure 5-3 shows relative voltage signal transfer, using a one-volt input, as a function of the ratio of the two

Voltage transfer

R_s (ohms)	R_L (ohms)	V_o (volts)
100	1	0.0099
100	10	0.091
100	100	0.5
100	1,000	0.91
100	10,000	0.99

Power transfer

R_s (ohms)	R_L (ohms)	P_{out} (mW)	P_{total} (mW)	P_{out} / P_{total}
100	1	0.1	9.9	1
100	10	0.81	9.1	8.9
100	100	2.5	5.0	50
100	1000	0.83	0.9	92

Figure 5-3 Maximal voltage transfer occurs when load resistance R_L greatly exceeds R_S. Maximal power transfer occurs when source and load impedances are equal.

resistances, where R_S (source or output resistance) is equivalent to R_1 in Figure 5-2 and R_L (load or input resistance) corresponds to R_2.

In cases in which we wish to transfer power, as in driving a mechanical loudspeaker, we must consider the transfer of current as well as the transfer of voltage because power is the product of voltage and current (Equation 5-8). Maximal power transfer occurs when both source and load impedances are equal. The column P_{out} shows that maximum power is delivered to the load R_L when R_S equals R_L.

The voltage divider circuit, when including frequency-sensitive passive elements (capacitors or inductors), will act as a **filter** – a circuit that allows some frequencies to pass while others are attenuated. Filters in use today rely mainly on resistors and capacitors. Inductors are less frequently used because they are physically large in an era of tiny surface-mount electronic elements, susceptible to electromagnetic interference, and unnecessary in many applications. Inductors can impart a characteristic sound that is sometimes desired and are again becoming popular as a way of intentionally altering the sound of some audio devices. The frequency above (or below) which attenuation occurs depends on the value of resistance and capacitance (or inductance).

Figure 5-4 shows a **low-pass filter**, in which low–frequency signals pass unattenuated. As the signal frequency increases, the capacitive reactance decreases. At the frequency at which the capacitive reactance just equals the resistance, the output signal is reduced by $\dfrac{1}{\sqrt{2}}$ (–3 dB). This frequency is known as the **corner frequency** of the filter and is determined by:

$$f_0 = \frac{1}{2\pi RC}$$

5-9

It might appear that the amplitude of the output signal at the corner frequency should be half the input and therefore the output should be down

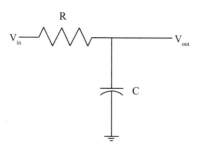

Figure 5-4 Low-pass filter.

by 1/2, or −6 dB, but there is another consideration: the capacitor has an effect on the phase of the signal as well as its amplitude. As the frequency of the signal increases, the time response of the capacitor begins to shift the phase of a sine wave signal as it flows through the capacitor. We must use a vector description of the circuit, one that deals simultaneously with the amplitude and phase of the signal. When the vector description of the impedances is used, the magnitude part of the vector sum is:

$$Z_{total} = \sqrt{(R^2 + X_C^2}\qquad\qquad 5\text{-}10$$

So, substituting in the voltage divider equation, we get:

$$V_{out} = \left[\frac{R}{\sqrt{R^2 + X_C^2}}\right] V_{in}\qquad\qquad 5\text{-}11$$

So at the frequency where $X_C = R$,

$$\frac{V_{out}}{V_{in}} = \frac{R}{\sqrt{2R^2}} = \frac{1}{\sqrt{2}} \Rightarrow 20\log\left(\frac{1}{\sqrt{2}}\right) = -3\text{dB}\qquad 5\text{-}12$$

A similar analysis can be applied to the **high-pass filter**, shown in Figure 5-5.

In this case, the capacitor effectively blocks DC and low-frequency components, causing most of the input voltage to drop across the high capacitive reactance. As the frequency of the signal increases, the impedance of the capacitor decreases and more of the input voltage appears across the resistor.

Frequently, an intuitive understanding of the operation of circuit elements is as helpful as a complete engineering analysis. Most circuits can be understood on a superficial basis, because one is not trying to design

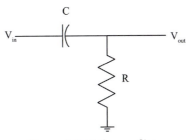

Figure 5-5 High-pass filter.

a circuit but simply appreciate what a circuit is doing to the signal in a general way, for example, boosting high frequencies, mixing signals, and buffering or impedance matching. Often, a circuit can be understood by using Ohm's law and the R–C circuit theory introduced earlier. As users of audio devices, we are not concerned as much with engineering precision as we are with understanding general circuit function.

Conceptually, capacitors have high impedance for low frequencies and inductors have high impedance for high frequencies. Exactly what we consider to be high and low frequencies depends on the relative values of the circuit elements. For a more precise quantitative understanding of passive device circuits and sinusoidal signals, the use of **phasors** – complex number descriptions of impedances – are in order.

As mentioned previously, the impedances of capacitors and inductors can be treated like resistances, but with an added dependence on the frequency of the signal not present in resistors. Because reactance is a function of frequency, the exact equivalent resistance changes with frequency, but for a given frequency, has a fixed value. We may treat all impedances similarly if we have a way of dealing with the frequency-dependent nature of reactance, which we represent using complex numbers (designated by bold type). For example, the impedance of a capacitor is given by:

$$Z_C = \frac{1}{j\omega C} \qquad\qquad 5\text{-}13$$

where $j = \sqrt{-1}$ and $\omega = 2\pi f$.

The impedance of an inductor is:

$$Z_L = j\omega L \qquad\qquad 5\text{-}14$$

If we use these expressions for the impedances of reactive circuit elements, circuit analysis is simplified.

Simple series and parallel combinations of impedances are often encountered in useful circuits like equalizers. By selecting the arrangement and values of these components, we can make frequency-selective circuits to modify the frequency composition of a signal by attenuating some frequencies more than others. (Note that using passive components alone, we can only attenuate and not amplify: amplification requires active circuit elements that often follow these passive circuits.) The high-pass and low-pass circuits are the simplest of these circuits: using more impedance elements in clever geometries, we can make filters that somewhat attenuate frequencies above or below a corner frequency (shelving filters), affect only the frequencies within a certain range (band pass filter), and cut very sharp, narrow bands of frequencies (notch filter).

Capacitors and inductors take finite time to build and release their stored energy. This time leads to a phase shift in the sinusoidal components of a signal as a function of frequency. For single-pole filters, those containing a single energy-storing component, L or C, the phase shift is already $45°$ ($\pi/4$ radians) at the f_0 corner frequency (Figure 5-6). The phase shift begins two decades below the corner frequency. Multipole filters, formed by combining single poles, have sharper cutoff slopes of 6 dB/octave per pole. The more complex filters can have resonances and introduce ripples in the frequency response magnitude as well as shift the phase of signal components. You will learn more about filter phase shifts when we discuss equalizers in Chapter 7.

Using inductors and capacitors together, we can take advantage of the phenomenon of **resonance**, as the circuit elements exchange energy back and forth at the characteristic resonant frequency, allowing oscillators and sharp filters to be created. A parallel arrangement of capacitor and inductor will exhibit low impedance at high and low frequencies. Capacitors have low impedance at high frequencies and inductors have low impedance at low frequencies, allowing high and low frequencies to pass the parallel circuit unhindered. At the resonant frequency, the frequency at which the inductive reactance and capacitive reactance are equal, the parallel impedance is very high. For series arrangements of capacitor and inductor, the

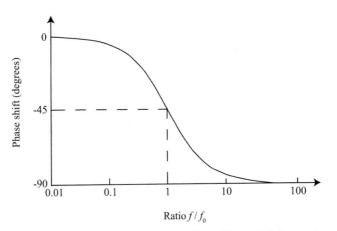

Ratio f/f_0

Figure 5-6 Phase shift for single-pole low-pass filter. Phase shift begins two decades below cutoff frequency f_0 and continues to increase up to two decades above f_0.

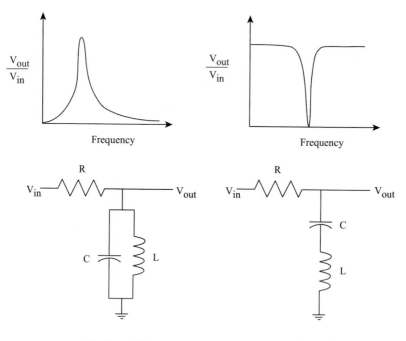

Band-pass filter   Notch filter

Figure 5-7 Resonant circuits result from capacitors and inductors passing stored energy back and forth.

opposite is true and the series impedance drops dramatically at the resonant frequency. When combined with a resistor in a voltage divider circuit, band–pass and notch filters can be created (Figure 5-7).

With the concepts of how impedances behave in series and parallel, we can begin to analyze and understand these more complicated circuits.

In **series**, impedances simply add linearly. In **parallel**, they combine as:

$$\frac{1}{Z_{total}} = \frac{1}{Z_1} + \frac{1}{Z_2} + \dots + \frac{1}{Z_n} \qquad 5\text{-}15$$

For two parallel impedances:

$$Z_{total} = \frac{Z_1 \times Z_2}{Z_1 + Z_2} \qquad 5\text{-}16$$

By using the complex number representation of the reactive imped-ances, we can calculate the behavior of useful circuits directly. But even without employing the mathematical analysis, we can grasp the function of

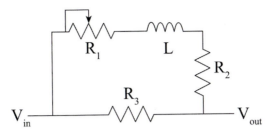

Figure 5-8 Low boost shelf filter. This circuit attenuates low frequencies less than higher frequencies.

a circuit by considering the relative contribution of each element's impedance as it relates to the whole circuit. Consider the circuit of Figure 5-8, a useful low-frequency tone controlling circuit.

The elements R_1, L, and R_2 provide a parallel path to R_3. When the parallel pathway has low impedance relative to R_3, it makes the total impedance low and the signal is passed through with little attenuation. Because for high frequencies X_L is large relative to R_3, the circuit will behave as if R_3 is the only element in the circuit at high frequencies. At low frequencies, X_L is low enough to ignore, and the total impedance of the circuit will be determined by R_1 plus R_2 in parallel with R_3. Because R_1 is adjustable, it will determine how much the parallel path shunts (bypasses) R_3 and therefore determines the overall impedance, with R_2 setting the minimum parallel impedance when R_1 is shorted. For low frequencies, the total impedance is adjusted by R_1; for high frequencies, it is determined by R_3. This rule gives us a useful filter: a low shelf where every frequency below a corner frequency is increased an adjustable amount relative to the higher frequencies. It should be remembered, however, that no gain is available in such a circuit and that we are actually attenuating the "boosted" frequencies less. Without considering the values of the components, we cannot tell exactly where the line between low and high frequencies will occur, but we can determine the general function of the circuit by inspection, keeping in mind the basic operation of the passive electronic elements.

ACTIVE ELECTRONIC DEVICES

As we have seen, passive electronic components can be used to alter the frequency content of signals but cannot amplify, or increase, the signal level. For this, we must rely on active devices, ones that can convert large applied DC voltages into amplified versions of smaller input voltages. These devices fall mainly into two categories: vacuum tubes and solid-state devices based on silicon or similar semiconducting materials. Each is able to use a small voltage or current to control a larger voltage or current, thereby producing amplification. The physical processes involved in these alternative systems are different, but each has advantages and disadvantages that allow us to choose one or the other for a particular task based on these relative strengths and weaknesses.

Vacuum or **thermionic tubes** were the earliest devices allowing amplification. They are somewhat like a modified lightbulb: an evacuated glass envelope with several metal electrodes. The **cathode** electrode is heated by electric current, through a filament, until it is hot enough to emit electrons. By placing a large positive voltage between the cathode and a second electrode, the **anode** or **plate**, a current can be made to flow as the free electrons are attracted to the plate. If a third electrode, the **control grid**, is placed between the cathode and the plate and made slightly negative relative to the cathode, a small voltage can be made to control the current from cathode to plate, the plate current. When the plate current is made to flow through a fixed resistance, a larger voltage proportional to the grid voltage can be produced. This three-electrode tube is called a triode (see Figure 5-9). Further refinement of the triode created the tetrode and pentode, with additional electrodes added to improve some aspects of the system. The tetrode adds a **screen grid** between the grid and plate that reduces the capacitance of the tube and makes the plate current less dependent on the plate voltage. The pentode also adds a **suppressor grid** connected to the cathode between the screen grid and plate to address the secondary emission of electrons that occurs as the plate current electrons strike the plate and dislodge stray electrons.

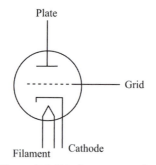

Figure 5-9 Triode vacuum tube.

Vacuum tubes share the advantages and disadvantages of high-impedance devices: they have inherently high input impedances and can produce large linear amplification, but they are plagued by thermal noise and capacitance issues and are vulnerable to mechanical vibration pickup ("microphonics") and air leakage. They also require external power supplied at hundreds of volts DC and must be heated to incandescence. Tubes can produce clean amplification if they are used in proper circuits and with careful shielding. When overdriven, they distort by current limiting rather than by voltage limiting at the output, producing harmonics, but without the relative harshness of abrupt voltage-limited clipping. When a device is current limited, the output voltage drops off smoothly as the current limit is approached, generating even-order harmonics. Voltage limits are reached abruptly, resulting in odd-order harmonics. Many tube amplifier circuit designs employ transformers to match the desired input and output impedances to the high voltages and impedances of the tube circuits, and these components also contribute to the classic sound of favored tube amplifiers.

A newer amplification device has revolutionized electronics and spawned the age of computers: the transistor. Unlike vacuum tubes, transistors can be ultraminiaturized, require little applied power, and are practically made of sand. These **solid-state devices** are based on the behavior of materials known as **semiconductors**. Metals are easily able to conduct electricity; they do so by releasing loosely bound outer electrons freely, allowing current flow as the electrons move in response to external electric fields. Insulating materials have no such electrons available and hence do not allow current to flow. Materials such as silicon and germanium that possess a limited number of free outer electrons and exhibit an intermediate ability to conduct current, are known as **semiconductors**. Solid-state devices all rely on the same basic structure: junctions of semiconducting layers, each containing specific chemical impurities (called **dopants**) that confer either a net surplus or deficit of electrons. The areas with surplus electrons are designated n (negative) and the layers lacking electrons p (positive). Places where electrons are missing in p-type material are considered to behave as mobile positive charges and are called **holes**. Both electrons and holes diffuse freely through the semiconductor material, and their movement is influenced by electric fields as well as by their concentration gradients. Solid-state devices are constructed by joining p and n regions of semiconducting materials in various combinations, leading to different behaviors that result from the physical construction and doping materials that are added to form the different devices. Joining one n and one p layer creates a **diode** junction, which passes current easily in one direction but not in the other. Stacking three layers in alternating fashion creates a **bipolar junction transistor**.

A diode is constructed from two adjoining areas of doped semiconductor, one p–type and one n–type (Figure 5-10). When a voltage is applied that makes the p area positive relative to the n region (forward bias), current flows and the diode conducts. When the voltage is reversed (reverse bias), the surplus electrons and holes are pulled away from the junction and little current flows. The I-V description of the diode is shown in Figure 5-11.

I_S is the saturation current, which is determined by the construction of the diode; q is the electric charge (A·s); v is the voltage (positive on the p side); k is Boltzmann's constant ($1.3806503 \times 10^{-23}$ m^2 kg/s^2 K); and T is temperature (K). This relationship results in very small current flow when v is negative, essentially I_S, which is in the nanoamp range. As the voltage

Diode

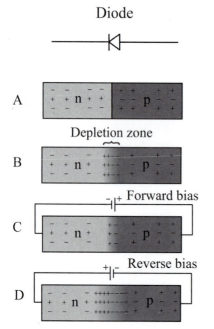

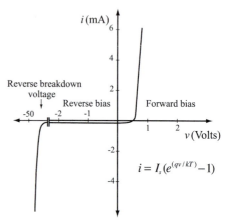

$$i = I_s(e^{(qv/kT)} - 1)$$

Figure 5-10 Solid-state diode and p-n junction. The depletion zone is created by a diffusion of charges without external voltage. Applying forward bias voltage causes the junction to conduct.

Figure 5-11 Diode I-V relationship plotted.

becomes positive, the junction begins conduction and at around 0.5 V, current increases dramatically for very small increments in the voltage. This nonlinear I-V behavior of a p-n junction differs from the linear behavior of resistors, whose impedance (I-V plot slope) is independent of the applied voltage, and therefore the diode does not obey Ohm's Law.

Although the theoretical description of p-n junction behavior does predict real-world behavior, it is an incomplete description of real diode in-circuit performance, especially with regard to complex signals like music. Fortunately, a simplified conceptual understanding of diodes is sufficient to understand their uses in audio equipment. One significant deviation from ideal behavior occurs with large reverse-bias voltages, which cause the diode to begin conducting even though the junction is reverse-biased. This result is known as the **breakdown voltage**; its value is determined by the physical construction of the diode. Though this occurrence can lead to thermal damage, it can also be used to generate a voltage reference as long as the power-dissipating capability of the diode is not exceeded. Special diodes known as Zener diodes are used in this mode to provide voltages that don't change as the current through them changes – generating stable voltage references in power supplies, for example. Diodes are used to rectify the AC voltage from the power line in DC power supplies. A frequent use for diodes in audio circuits is to rectify AC voltages in order to generate DC control voltages derived from signal amplitude envelopes, for use in compressors and expanders as well as in visual amplitude displays. The LED is a special variety of diode that emits photons of light in response to current flow and is of no small importance in the metering of audio signals.

By constructing two diodes back to back and sharing a central electrode, a **transistor** is created. The behavior of this device depends on the mobility of the free charges that can diffuse through the semiconductor and respond to the influence of electric fields that are created at the junctions between the p and n areas. Because the free electrons and holes are constantly in motion powered by thermal energy, they diffuse randomly through the silicon like ions in solution, even with no externally applied voltage. At the p-n junction, the electrons and holes are able to cross the junction, following their concentration gradients. The result is an equilibrium distribution of electrons and holes counterbalanced by the resulting electric field generated at the p-n junction by the separation of charges. This electric field is only present in the immediate vicinity of the p-n junction and does not affect the bulk of the semiconductor; however, charges that happen to drift close to the junction are swept across it by the electric field. Because electron and hole pairs are continually combining, and neutralizing their separated charges, as well as being generated anew by thermal energy, there is significant random movement of charge in the materials even without external voltages – a potential source of noise in active circuits.

In the most common transistor type, the bipolar junction transistor (BJT), the emitter is doped more heavily than is the base or the collector. In NPN transistors (Figure 5-12), this excess doping provides extra electrons that diffuse toward the emitter-base p-n junction. When a positive voltage is applied between the base and emitter, the B-E junction is forward biased and electrons flow into the base. Electrons enter and diffuse around the base as minority carriers, as holes are the majority carrier in the p-type base. The base is made small so that all of the electrons injected into the base by the emitter don't just combine with holes. When the collector is made positive relative to the base, the junction is reverse-biased, but electrons may still be swept across the base-collector junction by the electric field. Most of the emitter current flows out the collector lead, with only 1% or so of this current flowing in the base lead. If the base current is controlled externally (as by a signal voltage), the collector and emitter currents will be amplified versions of the base current and we have a method of increasing signal levels. PNP transistors work similarly but with reversed polarity: holes are emitted instead of electrons.

The bipolar transistor is often considered to operate as a current device: the current into the base controls a larger current flowing from collector to emitter, which is an analog of the base current amplified by the current gain (β or h_{fe}) of the transistor. As we have seen, however, it is the internal electric fields that most directly influence the device behavior. Transistors have inherently low input impedance at the base, unlike the high input impedance at the grid of a vacuum tube. The output voltage depends on the size of the collector resistor used in the circuit: the transistor's output is a current that

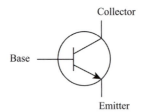

Collector

Base

Emitter

Figure 5-12 Bipolar NPN transistor.

creates a voltage drop in the load resistance. If a resistor is placed in series with the collector, the voltage at the collector is an amplified inverted version of the input voltage; this is an inverting voltage amplifier. If the resistor is instead placed in the emitter lead, the circuit can produce more current gain, but there is no voltage gain: this circuit functions as a noninverting impedance converter, or buffer. The transistor can also be connected as a current gain device, as it often appears in power supply circuits in which the base current is regulated to control the larger collector/emitter currents. Like diodes, transistors are nonlinear devices: they require a minimum voltage input to force the transistor into conducting current. There is a linear range that can be used for audio amplification if the input voltage is biased, or added to a fixed DC voltage, to center the input in the linear region of the I-V plot. When used in properly designed and constructed circuits, transistors can give clean, low-noise, high-gain performance in microphone preamplifiers and many other critical audio devices. They can also be used to create larger circuits contained on a single silicon substrate, making functional circuits like operational amplifiers (a standard integrated circuit amplifier device, known as op-amps), voltage-controlled gain devices, level detectors, oscillators, and filters as well as incredibly complicated digital circuits that use transistors as logic switches.

Another type of transistor often encountered in audio circuits is the **field effect transistor** (FET). FETs can be either J-FET or MOS-FET types; despite being constructed differently, both make use of the internal electric field to control the flow of charge through the device – hence the name. The junction FET (J-FET) shown in Figure 5-13 is constructed with the source and drain leads connected to opposite ends of a channel of one semiconductor type embedded in a bulk surround of the opposite type. When a voltage is applied between the source and the surrounding semiconductor (the gate), the electric field (e-> in Figure 5-14) at the p-n junction causes the depletion of carriers in the channel and thus acts to control the flow of charge through the channel, changing its resistance as a function of the gate voltage. Only the majority carrier in the source/drain material moves in the FET; the BJT operates with movement both majority and minority carriers at the p-n junctions. The drain-source resistance can vary from nearly infinite down to tens or hundreds of ohms, a very useful range for electronic switching and gain control. The p-n junction must never become forward biased in the FET or the gate will begin to conduct current as the diode junction switches on; therefore, the gate voltage must remain negative relative to the most negative end of the source/drain

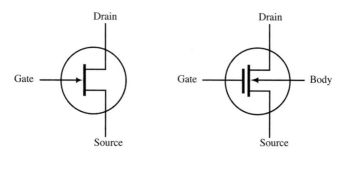

Figure 5-13 J-FET and MOS-FET (n-channel).

channel. By placing a resistor between the drain and the power supply voltage and grounding the source through a small resistance, a small gate voltage can be amplified as the drain–source current flows. This amplifier circuit does not invert the input signal. The FET is considered to be a voltage input device: the gate voltage controls the drain–source current and the gate provides a high–resistance input as well.

In the MOS ("metal–oxide–semiconductor") type of FET (Figure 5-15), two areas of the same type are slightly separated, embedded in a substrate consisting of the opposite type. For an n–channel MOS-FET, the device is

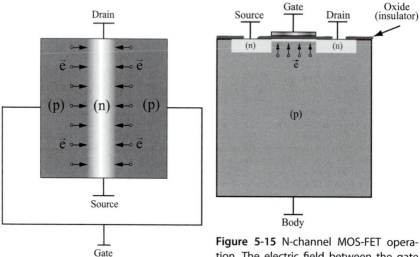

Figure 5-14 N-channel JFET operation where e indicates the electric field.

Figure 5-15 N-channel MOS-FET operation. The electric field between the gate and body creates an n-channel between the source and drain.

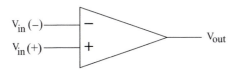

Figure 5-16 Operational amplifier (op-amp). The output is the amplified difference between the inputs.

constructed on a substrate of p–type material into which are doped a pair of separated n channel areas, the source and drain. The substrate surface between the two n areas is layered with an insulating oxide layer and then a metal electrode, the gate. Because the gate is insulated from the semiconductor materials, the MOS-FET gate has very high input impedance. The gate electrode and the substrate semiconductor beneath the insulating oxide layer form a tiny capacitor, on the order of 1 pF, which results in the accumulation of charges in the substrate area under the gate between the source and drain. The voltage at the gate thereby causes a channel of conduction between the source and drain to be formed by electrons attracted to the positive charges on the gate in the case of the n–channel FET. The gate voltage acts to control the flow of current from source to drain by altering the number of available carriers induced to join the conduction channel. Because of the very high input impedance of the MOS-FET, relatively small static charges reaching the gate can damage the device, making static protection necessary to prevent damage to these devices.

FETs are used where high input impedances are desired and also in power amplifiers and in digital circuits. The FET can be used as a variable resistance device as well as an amplifier or switch; hence FETs are commonly used as digitally controlled audio switches, providing a way of switching audio signal routing and muting under software control, and in compressors and gates as voltage-controlled variable resistance elements. FETs are useful as preamplifiers in part because they have high input impedances that are desirable for amplifying signals from high–impedance devices such as piezoelectric instrument pickups, where they function as impedance converters or buffers.

The availability of different amplification devices allows for a variety of circuit designs to accomplish a particular audio task. How a device will sound depends on many smaller influences as well as circuit interactions not always considered in our first approximation descriptions of the circuit elements. For example, the physical construction of a device may affect the sound it produces by creating small, stray capacitances and inductances in high-impedance circuits that cause instability at high frequencies. There are subtle and hard-to-define aspects of audio devices that may improve or degrade sound quality performance and resist an easy scientific analysis. There is an element of art in the design of audio circuits that comes with experience: both circuit design and implementation can contribute to potential differences in sound between two similar-looking pieces of gear. With the development of large, integrated circuits – entire functional circuits on a single chip – it has become easy to use off-the-shelf components to build equipment very inexpensively. Though more carefully crafted devices benefit from using fewer individual transistors and carefully selected circuit elements, simple audio circuits using an op-amp have made it possible to create audio devices that are easy to design and can sound quite good.

The operational amplifier was originally designed for use in analog computers, where circuits were created to do real-time computation using analog voltages to solve complicated problems. The operational amplifier, or op-amp, is a high-gain amplifier with two inputs that can be used to add (noninverting) and subtract (inverting), or it can be used as a linear amplifier whose characteristics are programmed by a network of attached components. Because the output is the amplified difference between the two input voltages, the amplifier is known as a **differential amplifier**. This feature proves to be particularly important in high-quality audio applications, as it can be used to reduce interference in long audio wiring runs. The flexibility of the op-amp makes it a tempting choice for general-purpose designs, and it is found in audio equipment of all performance levels. Op-amps can be made using bipolar transistors alone or in conjunction with field effect transistors, offering a range of possible configurations of these devices and allowing op-amps to be optimized for a range of different applications. Although many op-amps are integrated circuits, op-amps that are custom-made from selected transistors and other components and sealed in plastic modules are found in some high-performance audio equipment.

The ideal op-amp device is assumed to have an infinite input impedance and infinite gain. It is further assumed to have no current flowing into or out of either input and to have no voltage difference between the inputs. Analyzing two basic op-amp circuits will show how these assumptions make using op-amps easy. A resistor connected from the output back to the inverting input and an input resistor also connected to the inverting input forms an **inverting amplifier** (Figure 5-17). When the noninverting input is connected to ground, it causes the inverting terminal to act as

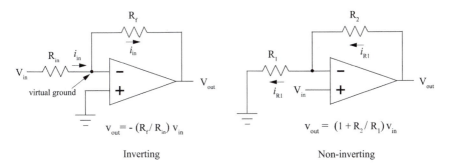

Figure 5-17 Inverting and noninverting op-amp circuits. Positive V_{in} causes current flow shown.

a **virtual ground**. Because there is infinite impedance between the inputs, no current is able to flow between them, and therefore by Ohm's law no voltage difference can exist between them. Current does flow in the input resistor and must flow out somewhere, according to Kirchhoff's law. Because it can't flow into the op-amp input, it must flow through R_f, generating a voltage drop across it that appears at the op-amp output. According to Ohm's law, the resistance of R_f determines the output voltage and therefore the gain of the circuit. Because the inverting input is at ground potential, the output voltage is negative when the input voltage is positive, due to the direction of current flow in R_f.

Because the inverting input is held at ground potential, connecting multiple input resistors to the inverting input sums the input currents with isolation between the sources. This phenomenon is the basis of an audio mixer. The input currents will all flow through R_f, but there will be no interaction of the sources because they all feed into the virtual ground voltage that does not change, no matter how much current flows through the circuit. The input impedance of the inverting circuit is just the resistance of R_{in}.

In the noninverting configuration of Figure 5-17, a portion of the output voltage is fed back to the inverting input through the voltage divider of R_2 and R_1. As before, the voltage at the noninverting input is reflected at the inverting input. The signal voltage at the inverting input causes a current flow in R_1. That current i_{R1} again must flow through R_2, but here the current in R_2 is in the direction that drives the output voltage positive for a positive input voltage. The output voltage is the total voltage drop across both resistors. The input impedance of the noninverting amplifier is very high.

Although in reality the ideal conditions we assume are not strictly met, they are often close enough to allow simple implementation of op-amp audio circuits, especially using the audio-optimized op-amps now readily available. Figure 5-18 shows some common op-amp circuits and how the arrangement of external circuit elements determines the overall circuit behavior.

In real op-amp integrated circuits (ICs), there are deviations from the ideal behavior that must be considered in designing audio devices. The

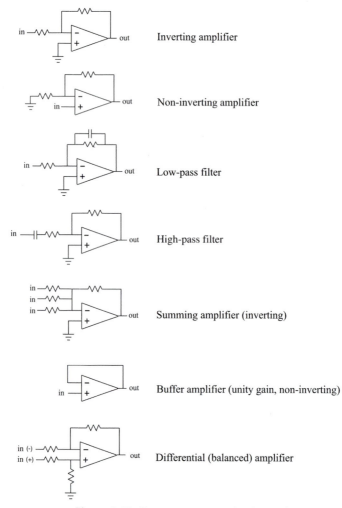

Figure 5-18 Common op-amp circuit topologies.

inputs generate small but finite input currents and offset voltages. Though these may be disregarded in some applications, they must be considered when dealing with high-input impedances in which the small leakage currents may generate significant voltages. The input offset voltages must also be considered in very high-gain applications. Further, any time lag in the feedback circuit may degrade the stabilizing effect of the negative feedback at high signal frequencies. The rate at which the output voltage can change is limited, and the effect of this limit on large, high-frequency signals is known as **slew-rate limiting**. This effect is different from the overall high-frequency limits of the amplifier and applies only to fast, high-amplitude output swings. Thus the op-amp should be carefully selected for the type of application in order to achieve the highest performance. Op-amp integrated circuit designs have been greatly refined over the years, and many op-amps available now are optimized for audio performance.

SUGGESTED READING

Horowitz, P., & Hill, W. (1989). *The Art of Electronics* (2nd ed.). Cambridge University Press. ISBN: 0521370957. Good general and practical text on electronics. Chapters 1–7 discuss analog topics.

Jung, W. G. (1986). *IC Op-amp Cookbook* (3rd ed.). Howard W. Sams & Co. ISBN: 0-672-22453-4. A how-to book on filter design.

Jung, W. G. (2004). *Op-amp Applications Handbook (Analog Devices Series)*. Newnes. ISBN: 0-7506-7844-5. Op-amp circuit design.

Microphones

Contents

Microphones are the most critical element in the recording chain. Every sound not synthesized purely electronically must be transduced through a microphone in order to be recorded. There is a bewildering array of available microphones, and each may be optimally suited to a slightly different application. One of the most important areas of knowledge for a recording engineer is the proper application of microphones: selection and placement both require a solid understanding of how microphones work to obtain the best possible results. This knowledge is ultimately achieved through experience, but understanding the properties of the various microphones helps accelerate the process.

Microphones may be classified both according to the physical method of transduction (Figure 6-1) and to their spatial sensitivity pattern. Although these behaviors are related, the relationship is complicated and requires a thorough examination of the different physical transducer types. The main types of transducers used for recording are **dynamic**, whereby a conductor

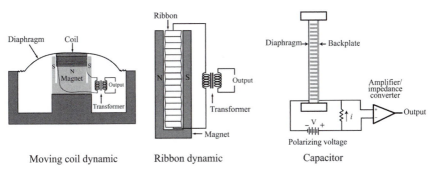

Figure 6-1 Microphone transducer types.

The Science of Sound Recording
ISBN 978-0-240-82154-2

moves within a magnetic field in response to the forces applied by incident sound waves, and **capacitor** (also known as condenser), whereby one plate of a capacitor moves relative to the second fixed plate. All of these devices convert air vibrations into proportional electrical voltages. There are variations on these themes, and some other types are beginning to be developed, but we can begin by examining the standard dynamic and capacitor types.

DYNAMIC MICROPHONES

Dynamic microphones depend on the principle of induction, in which moving a conductor within a magnetic field induces a current in the wire. In the case of **moving–coil dynamic microphones**, the conductor is a coil of very fine diameter wire situated within a magnetic field and attached to a diaphragm in contact with the air. As the pressure varies, the diaphragm moves in response to the changing force applied by the moving air. The coil produces a small voltage as it moves in the fixed magnetic field. This voltage is fed, usually through a transformer, to an external amplifier optimized for low input impedance and high gain. The output voltage is proportional to the velocity with which the coil moves through the magnetic field, making the microphone inherently sensitive to the air particle velocity as shown in Equation 6-1:

$$e(t) \; = \; Blu(t) \qquad\qquad 6\text{-}1$$

where $e(t)$ is instantaneous output voltage, B is the magnetic field strength, l is the length of the conductor, and $u(t)$ is the instantaneous velocity of the conductor. Although conceptually simple, the implementation of the dynamic microphone is not so straightforward. The mass of a coil of wire is not negligible, so the construction of the element requires special care to make sure the element can move easily enough to allow the small air pressure variations to produce a measurable voltage at all audible frequencies. There are also acoustical considerations necessary to produce a consistent output level for sounds originating from different directions and at different frequencies. This need often leads to complicated acoustic labyrinths built into the housing of the microphone in order to control the frequency response and directional sensitivity.

A second type of dynamic microphone is the ribbon microphone, in which a small ribbon of corrugated metal is placed within a magnetic field and allowed to move in response to air movement from both the front and rear of the ribbon. Ribbon microphones have lower conversion efficiencies and produce lower output voltages for a given sound level, but they are lower in mass than a moving coil element and can produce better high-frequency transduction. Because both the front and back of the ribbon are in contact with moving air, the ribbon microphone is sensitive to sounds from the front and rear but not to sounds directly from the sides. Forces from the side of the ribbon are ineffective at moving it, so no output is produced. This design produces a distinctive bidirectional or figure-eight polar sensitivity pattern, although the characteristic pattern may be altered using acoustical techniques to control the access of the air to the ribbon. Ribbon microphones were pioneered by RCA in the early 20th century and were used in many of the well-known recordings of that period as well as for radio announcing. Using new ribbon materials and much stronger rare-earth magnets, the ribbon microphone has achieved renewed popularity due to better durability and higher output levels than older designs.

CAPACITOR MICROPHONES

The concept of **capacitor element** operation is quite simple: a capacitor is formed by a thin conductive membrane stretched by a supporting ring placed close to a second fixed plate with an air dielectric between them. Because $v = q/C$, holding v constant and changing C forces a reciprocal change in q, which is by definition current. The current is made to flow through a very large resistor and the resulting voltage drop is buffered and amplified to become the output of the microphone. The voltage produced by an omnidirectional capacitor microphone as a function of pressure is approximated by Equation 6-2:

$$e = \frac{E_0 a^2 P}{8 h T_0} \qquad 6\text{-}2$$

where e is open-circuit output voltage, E_0 is the polarizing voltage, a is the diaphragm radius (m), P is the pressure (Pa), h is the distance (m) from the diaphragm to the backplate, and T_0 is the diaphragm tension (N/m). The capacitance change is due to the change in h as the diaphragm moves relative to the backplate. Changes in h of from a peak of 10^{-6} m to as little as 10^{-11} m are sufficient to produce outputs. From Equation 6-2, we can see that the output voltage increases with increasing diaphragm radius a, with decreasing diaphragm-to-backplate spacing h, and with decreasing tension T_0. Of course the situation is more complicated because the tension is also adjusted to control the resonances that affect the frequency response of the microphone. The distance h must be sufficient to prevent the diaphragm from contacting the backplate, and the tension T_0 must also be high enough to prevent such contact. Different approaches to balancing these parameters lead to different-sounding microphones.

Unlike the dynamic microphone, the capacitor microphone requires an externally supplied source of voltage (E_0) to operate, so either a battery or phantom power is required to power the circuitry that converts capacitance changes into voltages. **Phantom power** is a clever method of applying external power by raising the two signal lines to a common DC voltage through the XLR cable driven by the preamplifier. The voltage is then extracted by the microphone circuitry while the DC potential is ignored by the capacitor- or transformer-coupled differential input of the preamplifier. The phantom power voltage is usually 48 volts DC and care must be taken not to turn it on or off while the microphone is audible.

Some capacitor microphones use a permanently charged plate instead of an applied voltage to polarize the capacitor: these are known as **electret capacitor elements**. Either the diaphragm itself or the backplate may be constructed of material containing a permanent electric charge. Electret capsules still require an applied voltage to power the internal electronics. The construction of the capacitor element involves art as well as science and the technical details of production are closely guarded secrets in many cases. The materials used play a role as do the design and manufacturing techniques; there are many variables in building a capacitor microphone, and the quality can vary dramatically for two similar-looking microphones. Because the mass of the diaphragm is important, the composition, thickness, coating, and tensioning of the diaphragm are critical. The tensioning of the diaphragm is something of an art, which makes careful assembly of high-quality capacitor capsules necessary and increases the cost of such capacitor microphones. The capacitor microphone can also be constructed using a radio frequency signal to extract the change in capacitance using a balanced AC bridge instead of a DC polarization.

Although the transducer types inherently behave as either pressure- or velocity-sensitive, changing the acoustical properties of the microphone can alter that behavior. Using both mechanical damping and acoustic pathways, the response of the sensing element may be adapted to allow dynamic and capacitor microphones to behave as either pressure or pressure-gradient microphones (explained in the following section) as desired. These treatments also allow the inherent resonances of the sensing elements to be tailored to a flatter response by introducing complementary resonances and damping the existing ones. The wide variety of microphones available displays the sonic effects of these varied design approaches and gives us the ability to choose a microphone well suited to each different application and budget.

SPATIAL SENSITIVITY

In addition to the type of microphone transducer construction, we also classify the way the microphone responds to sounds coming from different directions. A microphone may be equally sensitive to sounds regardless of direction, a type known as **omnidirectional**, or it may favor sounds from a certain direction over others, in which case we call it a **directional** microphone. It is very difficult to make a microphone that behaves perfectly; for example, omnidirectional types still have slight sensitivity differences to sounds at higher frequencies coming from some directions. All real microphone sensitivity patterns vary from theoretical predictions, and these imperfections contribute to the distinctive sounds of the microphones we use. Understanding how theory relates to practice is the goal of recording engineers who must use microphones to begin the recording process.

Despite the absence of an absolute relationship between transducer type and spatial sensitivity pattern, there is an aspect of microphone behavior that relates directly to spatial sensitivity. Microphones may be classified as to whether they are sensitive to absolute pressure, in which case they are necessarily omnidirectional, or to a pressure gradient from front to back, in which case they are directional. **Pressure microphones** are sensitive only to the pressure at the front of the capsule and have no way of determining from which direction the pressure wave originates. **Pressure-gradient microphones** produce an output proportional to the difference in pressure between the front and back of the capsule and therefore react differently to sounds from different directions. Pressure-gradient microphones often employ acoustic signal manipulation within the body to achieve the desired spatial sensitivity by delaying and attenuating the sound that strikes the microphone rear ports. Using the length and acoustic resistance of the path to the rear of the sensing

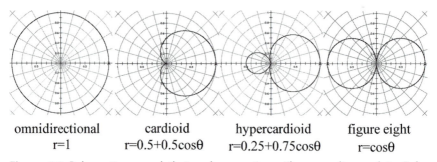

omnidirectional	cardioid	hypercardioid	figure eight
r=1	r=0.5+0.5cosθ	r=0.25+0.75cosθ	r=cosθ

Figure 6-2 Polar patterns and their polar equations. These are linear plots. Polar patterns are also sometimes presented as log plots.

element, the microphone can be constructed to exhibit the desired spatial sensitivity pattern.

The polar patterns shown in Figure 6-2 show the various spatial sensitivity patterns and the equations that describe them. Often called **polar patterns**, the radius r is a measure of the microphone's sensitivity at the angle $θ$ relative to the front of the microphone, which here is the positive x-axis. It can be seen that the overall spatial sensitivity is the sum of contributions from an omnidirectional (pressure) term and a bidirectional cosine (pressure-gradient) term. By varying the ratio of the terms, we can create any desired polar pattern. In fact, variable-pattern microphones accomplish this by combining two capsules, although not always the two mentioned: a back-to-back cardioid pair is often used, as it is easier to construct.

Unfortunately, the polar patterns presented by manufacturers may be somewhat misleading. Real spatial sensitivity patterns deviate from the ideal ones and often vary significantly with frequency. Off-axis response can seriously alter the sound of a microphone in real-world use, as in a room there are reflections of the direct sound picked up off axis and combined with the on-axis signal. The combination can radically alter the transduced sound quality through frequency-dependent reinforcements and cancellations. One of the major differences between great microphones and others is the smoothness of the off-axis frequency response. Because omnidirectional microphones have better off-axis response than directional microphones, they tend to produce less coloration in practical use.

An important characteristic of directional microphones is the ability of the transducer to select sounds coming from the on-axis direction and reject those coming from other directions. The nulls in the polar patterns indicate the direction from which little or no pickup will occur. The figure-eight pattern has very good rejection of sounds approaching from the sides and can be used to record a singer also playing guitar, for example, if the null of the guitar mic is aimed at the singer's mouth, surprisingly little vocal sound will appear in the guitar microphone. Because off-axis sounds are generally ambient sounds, the ratio of on–axis to off-axis pickup gives a measure of the microphone's ability to focus on the desired direct sound and reject ambient sound. This characteristic is known as the "reach" of the microphone, a measure of how far the microphone may be placed from the source while maintaining the preferred ratio of direct to reflected sound. It is often desirable to place microphones at some distance from an instrument, as such placement allows sound radiated from the whole instrument to be picked up more evenly. A simple measure of a microphone's directional selectivity is the distance factor (DF), whereby the microphone's selectivity is measured against that of the omnidirectional pattern. With the omnidirectional DF=1, the cardioid and figure-eight have DFs of 1.7 and the hypercardioid has a DF of 2; hence a hypercardioid can be placed twice as far from the source as an omnidirectional and maintain the same ratio of direct to ambient sound. Other ways to measure this characteristic include the random energy efficiency (REE) and the directivity factor. All give essentially the same information. Figure 6-3 shows the relative distances for equivalent direct-to-ambient pickup for several polar patterns.

Directional microphones all exhibit some degree of **proximity effect**, an increase in output level for low-frequency sounds as the distance from the source decreases. For cardioids, often used for close vocal recording, this effect means that the closer you sing to the microphone, the more the lower vocal frequencies are boosted. Although the effect can be used artistically, it may also cause problems because it tends to increase the effect of breath

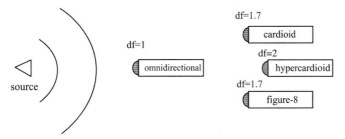

Figure 6-3 Distance factor showing relative distance from source producing the same balance of direct to reflected sound pickup for different polar patterns.

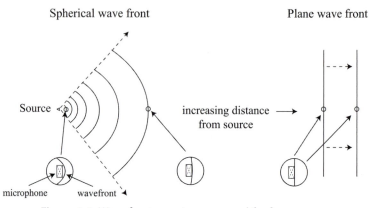

Figure 6-4 Wave front curvature near and far from source.

sounds and plosives like the /p/ and /b/ phonemes. Many directional microphones have selectable high-pass filters built in to reduce this effect when it is not desired. The cause of the phenomenon is explained by assuming that sounds originate as point sources and therefore propagate as spherical waves near the source (Figure 6-4). As each concentric sphere grows, the area through which a given angle of sound field energy radiates increases in proportion to the surface of the growing sphere, which increases with the square of the radius r (area $= 4\pi r^2$), here the distance from the source. The same amount of energy passes through larger and larger areas, rapidly decreasing the amount of energy per unit area. Far from the source, sound waves propagate nearly as a plane wave in which the energy flows through nearly the same area as the distance from the source increases. The pressure gradient is therefore greatest near the sound source.

   The total energy moving a directional microphone diaphragm is the sum of two contributing terms: pressure and particle velocity. In a spherical wave that occurs near a source, particle velocity falls off more rapidly with distance than does the pressure, and the two are out of phase. Far from the source where the wave is planar, pressure and particle velocity decrease equally with distance to the source and pressure and velocity are in phase. At low frequencies (long wavelengths), the particle velocity is normally low. A small decrease in distance close to a sound source results in a significant pressure-gradient increase between the front and rear of the microphone that increases the particle velocity, particularly for long wavelengths. This effect produces an increasing low-frequency boost as the source moves closer to the microphone.

The dimensions of a microphone affect the way the sensing element interacts with sound waves. Because the behavior of waves depends on their wavelengths relative to the size of the object with which they interact, we need to consider the capsule dimensions relative to the sound wavelengths we need to transduce. Most studio microphone element diameters are in the 12mm (1/2") to 25mm (1") range. There is a definite difference in the behavior of microphones, even within this relatively small range of diaphragm dimensions. Large-diameter elements – those with a diameter of 25mm or larger – begin to exhibit shadowing behavior at higher audible frequencies. Larger diaphragms also begin to exhibit more complex resonances that alter the output signal. For this reason, small diaphragm elements often produce better results in certain applications, notably when we want minimal coloration of the sound. In other situations, the inherent resonances of larger diaphragms can actually enhance the sound, as with many vocal recordings.

MICROPHONE SPECIFICATIONS

We must decide between bewildering possibilities when we choose a microphone for a specific task. Some valuable information can be hidden in those cryptic data sheets so thoughtfully provided by the manufacturer. Unfortunately, the critical information is frequently encoded into the specifications in different ways, making comparisons difficult (the more skeptical engineer might suspect that this is done on purpose). The main criterion for comparison would involve the microphone's **sensitivity**: how much signal is produced by a given SPL? Other considerations involve the amount of noise inherent in the microphone, the spatial sensitivity (polar pattern), and the overall frequency response of the microphone.

To compare the sensitivity of two microphones, their stated output must be measured relative to a standard pressure input. There are, unfortunately, several reference pressures commonly used, making a conversion necessary to get a direct comparison. The most directly useful sensitivity rating would be volts/dB SPL. What we frequently get is something like −75 dB re: 1V/mbar. A helpful conversion factor is:

$$10 \text{ dynes/cm}^2 = 10 \text{ mbar} = 1 \text{ pascal } (\text{N/m}^2)$$

$$= 94 \text{ dB SPL (all are units of pressure)}$$

In order to compare microphone sensitivity specifications, they must be converted to the same reference; only then can the relative output levels be directly compared.

Along with sensitivity, other considerations include the noise level generated by the microphone and the maximum sound pressure that can be transduced without distortion. The inherent noise in a microphone is often reported as "self-noise" or equivalent input noise. This is usually given as the sound pressure level that would be required to generate the observed output noise voltage. Smaller transducers usually have relatively more noise than larger ones. Most noise in dynamic microphones is generated by impedance-related random thermal processes; hence the low impedance of most microphones. Overloads in dynamic microphones are generated by the physical limits of the diaphragm, which is mechanically damped, allowing very high sound pressure levels (up to 140 dB SPL) to pass undistorted. Noise in capacitor microphones is mostly generated in the internal electronic circuitry. Overloads usually occur at the limits of the electronics' power supply voltage rather than due to physical excursion limits in the capsule. Even capacitor microphones can usually handle SPLs up to 130 dB, with many capable of handling SPLs up to 160 dB with the use of internal pads (attenuators).

There is a lot of discussion about the size of a microphone's diaphragm and its behavior with regard to low- and high-frequency transduction. The question is usually whether a small-diaphragm or large-diaphragm microphone is better for a given application. To understand performance differences between large- and small-diaphragm microphones, we need to consider how the sound pressure couples to the diaphragm. This knowledge will help determine how accurately the microphone converts the sound into an electrical signal.

We would prefer a microphone to act as a point source transducer so that its dimensions are negligible with respect to the wavelength of the sound pressure waves. When the wavelength of the sound approaches the dimensions of the microphone, their interaction begins to change. As the frequency of the sound increases, the dimensions of the microphone become closer to the wavelength of the pressure waves. This effect causes the microphone diaphragm to behave like a drumhead with several more complex modes of vibration than a piston. Generally speaking, small diaphragms will behave independently at higher frequencies than will larger diaphragms. Small-diameter (1/4" to 1/2") omnidirectional microphones deliver the most accurate transduction at high frequencies, so these are often used for critical recordings and research.

MULTIPLE-MICROPHONE TECHNIQUES

Although the great majority of modern recordings make use of complex multitrack systems and instrument overdubbing, there remains a school of thought that simple stereo recording is preferable, especially for live recording. There are many ways to accomplish this type of recording, several of which make use of just two microphones. As you will soon discover, however, it is not always as simple as putting two microphones in front of a musical group: the distance of each instrument as well as the individual instrument sound level outputs may vary too much to create a proper balance. Also, we are used to hearing stereo in a room with our two ears. Our hearing is designed to make use of two main cues about sound source placement: relative loudness and time of arrival. Sounds originating closer to one ear will be louder in that ear and will arrive sooner than at the other ear. In fact, other more subtle information also contributes to our perception of the sound field: phase relationships within the complex sound pressure signal that we hear can convey information about the height and front-to-rear placement of a sound source, interacting with the pinna of the ear in a way unique to each individual. Each auditory system has adapted to its own particular input filtering to produce the sensation of hearing. This adaptation is described by the **head–related transfer function** (HRTF), a mathematical model of the input filtering produced by sound waves interacting with the ear and head. Unfortunately, microphones do not "hear" the same way we do; no adaptation takes place except through the engineer's perception and experience. Consequently, recorded signals may not always convey the original sounds the way we would have heard them live when they are played back. The great challenge of stereo microphone technique is to bring to the listener a convincing image of the actual sonic event.

The key to capturing a convincing stereo sound field is similar to what allows us depth perception in vision: the overlap of sensory input from two separate sensors. Based on the cues we use to determine spatial placement of sounds, the way we select and position the microphones will determine how realistically we are able to recreate the spatial origins of the sounds we are capturing. Of course, in the real world there are confounding factors complicating the decision, like background noise, physical limits on where microphones may be placed, and limited time to try the alternatives. Considering the systems commonly used for stereo microphone placement

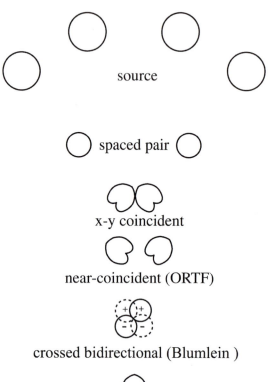

source

spaced pair

x-y coincident

near-coincident (ORTF)

crossed bidirectional (Blumlein)

mid-side

Figure 6-5 Common microphone orientations for stereophonic pickup. Distances from the source are not to scale.

will provide alternatives that can be used for a range of different recording challenges.

Figure 6-5 shows many of the common orientations using two microphones. Probably the most straightforward stereo placement is spaced pair referred to as **A–B placement**: two identical microphones are placed, at some distance apart, in front of the source. Omnidirectional microphones are frequently employed for this setup, although directional microphones can be used. This system captures both time of arrival and relative amplitude information, but if the microphones are spaced farther apart than our ears are, as is usually the case, the reproduced stereo field can be unnatural sounding. This approach might sound acceptable when listened to on speakers that are also placed far apart. The stereo separation reproduced depends on the distance between the microphones. A "hole" can be created in the middle of the stereo field if the microphones are too far apart. A third microphone is sometimes used in the middle of the spaced pair to prevent this effect, but undesirable cancellations may occur if this is not done carefully. Individual microphones (accent microphones) can also be placed near a soloist and combined with the spaced pair if necessary. When using spaced-pair recording techniques, it is helpful to consider the **3-to-1 rule** to minimize undesirable phase cancellations: the sound sources should be at least three times as far apart as they are from the microphones. (Obviously, this placement is not possible for a single point source.) Observing the 3-to-1 rule helps ensure that phase cancellations will be reduced to an acceptable level due to the inverse square law that governs the dissipation of sound intensity with increasing distance. The interaction of the microphones is important not only in stereo recording but also any time more than one microphone is used to pick up the same sounds, as in the case of multitrack studio recordings.

X-Y placement refers to the use of two closely placed microphones when the outputs are simply recorded and not in a matrix to produce the stereo sound. The angle between the microphone axes determines the apparent width of the stereo field. Angles between 90° and 135° are often used. Narrow angles emphasize the center of the sound field; wider angles create a wider image but may leave ambiguity in the center of the image. Many variations of this simple system of stereo microphone placement are possible using directional or omnidirectional microphones and either **coincident** or **near-coincident placement**.

When two directional microphones are placed together and aimed at an angle with respect to each other, a stereo recording can be created due solely to the amplitude differences because the time of arrival (or phase relationship) will be the same for both mics. (This placement does not work with perfectly omnidirectional microphones; however, most real omnis are not perfectly omnidirectional at all frequencies, as can be seen from their off-axis response at high frequencies.) This technique is also known as **single-point stereo microphone placement**. Coincident placement can be used to record a single instrument in stereo or for ensembles, but it may not be optimal for large groups because the far ends of the sound source may not be adequately picked up. When two bidirectional (figure-eight) microphones are used at a 90° angle, the system is known as a **Blumlein pair**. The Blumlein pair produces a very natural sound, but because the rear is picked up as well as the front, the placement is sensitive to sounds coming from the rear. This placement tends to work best close to the sound source in a good-sounding room without a restless audience. Also, sounds coming from the sides of the array will appear out of phase in left and right outputs, leading to potential problems if the stereo signal is mixed to mono.

A slight separation of a coincident pair can also yield a pleasing stereo image, providing the sound field is not too wide. The **ORTF** (Office de Radiodiffusion Télévision Française) has devised a method of separating two cardioid microphones 17 cm (6.7") at an angle of 110°. This system yields a good localization and depth of field because the capsules are close at low frequencies but adequately separated at higher frequencies to give some time-of-arrival information in addition to the relative level difference. Sounds arriving from far left and right may cause mono compatibility problems because of interference caused by the time of arrival difference, so the placement distance should be checked by listening in mono as well as stereo.

M–S (mid–side) technique refers to coincident placement of a forward–facing directional (middle or mid) and a figure-eight microphone (side) oriented at 90° with its front facing left so that the directional microphone faces the sound source. The outputs are processed in a matrix that produces sum and difference signals (M + S, M − S), which become the left and right signals (see Figure 6-6). On a mixer, the mid signal is panned center and the side signal is routed to two input channels with one panned left and the other inverted and panned right. By varying the mid to side mix, the stereo field width can be changed. This system has two major advantages: first, when the signals are combined in mono, there is a cancellation-free output, because the side mic cancels and only the mid signal is reproduced. Second, the stereo separation can be controlled by the

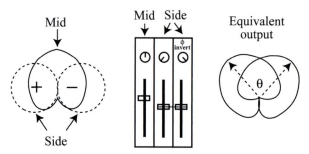

Figure 6-6 Decoding a mid-side recording. The mid microphone is panned center. The side-facing figure-eight microphone is fed to two channel strips. One is panned left and the other is inverted and panned right. The angle θ between the resulting phantom cardioids is determined by the ratio of mid to side levels.

matrixing operation, allowing the stereo spread to be changed after the recording is made.

A somewhat more complicated system that allows even greater post-recording manipulation of the sound is the **Ambisonic microphone**. It consists of four separate directional capsules mounted in the faces of a tetrahedron (three-sided pyramid) so that they aim at the odd-numbered corners of a cube: the elements are left-front up, right-rear up, right-front down, and left-rear down. These signals can then be matrixed to produce a wide range of simulated pairs and some ambience. It is not quite as simple as the M-S technique, however, and it takes a bit of experience (to say the least!) to understand what one is doing with this system. The Ambisonic microphone output can be recorded directly as B-format signals (called W, X, Y, Z) that can later be matrixed to produce a range of simulated stereo pairs. In B-format, the W signal is the pressure component equivalent to an omnidirectional microphone, and the X, Y, and Z signals are the equivalent of front, left, and upward-facing figure-eight microphones. Using the four signals, any pair of microphones oriented in any direction can be derived.

Binaural recording is a unique way of recording stereo in which a simulated human head, with microphones where the ears normally go, is used to record the signal and listening is done with headphones. This method can produce a convincing recording of a sound field. Unfortunately, there is one large drawback: it works poorly when reproduced on speakers. Several manufacturers make dummy head recording systems that include head, ears, and shoulders. Capacitor capsules are usually located in the ear canals. By placing the dummy in a hall, recordings that sound very much like "being there" can be made – provided that one listens on headphones. Recently, systems have been developed that use digital signal processing to render the effect on loudspeakers, but the system works only for a small listening area and requires the listener to remain stationary. A similar effect can be created using two spaced omnidirectional microphones by placing sound absorbing material between them. This technique, known as the **Jecklin disc**, places a 35 cm ($13^3/_4$") diameter disc covered with sound-absorbing material between two omnidirectional microphones spaced 36 cm ($14^3/_{16}$") apart.

The most convincing method of stereo recording is binaural, whereby a head with ears is approximated by the transducer system. This approach accurately reproduces both time of arrival and relative intensity cues but requires headphones for optimal reproduction. Any of the other techniques compromise some amount of potentially important information. So we are left to decide which cues are dominant in a particular situation. Choosing between a coincident pair and spaced omnis will often come down to trying both and finding out what works best for you in a given situation.

Similar principles governing microphone interactions are again imposed when we use multiple microphones to record in the studio. The phase cancellations we must consider when recording stereo are also possible when we use several microphones on a drum set. These undesirable effects may be clearly audible when an overhead microphone is soloed and then mixed with other microphones in the array, one at a time. If the sound changes drastically when the two microphones are mixed together, there are probably cancellations occurring. In this situation, we do have some options: moving the microphones or delaying them so the time of arrivals line up. Many preamplifiers have a switch to invert the signal that can also be used to find the best-sounding combination. Often, a small change in placement will make a large change in the interaction of the microphones, so some time spent perfecting the setup will pay off in the final sound quality.

Although impossible with analog recording, time alignment using digital recording and/or mixing is simple: we can delay the close-miked sources until they time-align with the more distant ones. For the drum set example, we use the overheads as our main source and delay the snare and kick mics so that they now align with the overheads. Because sound travels about 1130 feet/sec, we can delay the close-miked channels roughly 1 msec per foot of distance from the overhead mics. Although this trick sometimes tightens up the sound, it is by no means always necessary, and because all the microphones pick up some of all the other sources at different distances/delays, perfect alignment is impossible. This effect is why fewer microphones, properly placed, often deliver a more focused sound than a large array.

SUGGESTED READING

Gayford, M. (1994). *Microphone Engineering Handbook*. Focal Press. ISBN: 0-7506-1199-5.
    More technical coverage of microphone design.
Huber, D. M., & Williams, P. (1998). *Professional Microphone Techniques*. MIX Books. ISBN:
    0-872886-85-9. A more introductory discussion of microphones and their use.
Rayburn, R. A. (2012). *Eargle's Microphone Book* (3rd ed.). Focal Press. ISBN: 978-0-240-
    82075-0. Updated review of microphone designs and applications.

CHAPTER SEVEN

Analog Signal Processing

Contents

Amplifiers are the basic building blocks of analog signal processing devices. The narrow definition of amplifier is a "device that allows signals to be increased, either by increasing the voltage, the current, or both." More generally, though, **amplifiers** are the active electronic circuits that allow us to perform most of what we consider audio processing. In addition to simply combining and managing the amplitude of audio signals, there are special function amplifiers that offer the ability to selectively operate on separate frequency components of input signals independently. Other amplifiers have nonlinear gain structures that allow dynamic range processing. Mixing consoles are constructed of many specialized amplifiers, giving us control over amplitude, panorama, dynamics, and frequency content while combining and altering individual sounds. They can be made frequency sensitive and amplitude sensitive, creating control over spectral modification and dynamic range. Amplifiers drive the loudspeakers in the final stage of the recording process. We will also explain how to connect amplifier systems, avoiding some of the pitfalls that can degrade sound quality, so that an audio system may perform at its best.

The voltage signal created by most transducers is quite small – too small to do appreciable work. Because the ability to do enough work to leave a permanent magnetic particle polarization on a recording medium or cause a loudspeaker to move is a requirement, some way of increasing the power level of the signal is required. Electronic circuits are used to amplify the signal. This amplification requires an increase in the voltage and/or current

The Science of Sound Recording
ISBN 978-0-240-82154-2

capacity of the signal in a linear way. As you will see, there are applications for nonlinear amplifiers as well.

Devices capable of increasing the signal amplitude require the input of external electrical energy that the signal can be used to modulate (passive resonant circuits may appear to amplify signals, but only at the exact resonant frequency). These include thermionic devices (vacuum tubes) and solid-state devices such as transistors. Each such device uses the small voltage of the signal to control a larger voltage or current provided by a power supply. Though simple conceptually, the quality of amplification is dependent on perfect linearity and limited introduction of noise by the circuitry – conditions not always easily met. The design of the appropriate amplifier depends on factors determined by the device providing the signal as well as the device to which the signal is sent.

IMPEDANCE MATCHING

Figure 7-1 represents the conditions that exist when we connect two electronic devices. The source device is modeled as a voltage source in series with the impedance Z_{Source}, a circuit known as the **Thevenin equivalent** circuit. (A similar circuit using a current generator instead of a voltage source is the Norton equivalent.) The entire circuitry of any device can be considered as a single voltage source and impedance because this is what the output acts like to the connecting device's input.

In addition to the amount of voltage gain, one of the important characteristics of an amplifier is the impedance at the input and output. To transfer the voltage signal from one device to the next, the load (the input impedance being fed) must be larger than the source or output impedance by at least an order of magnitude. This requirement is especially important when an electromechanical device like a dynamic microphone is connected

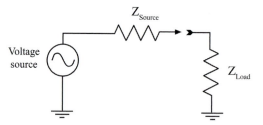

Figure 7-1 Load and source impedances form a voltage divider. Z_{Load} should be large relative to Z_{Source} for optimal voltage transfer between devices.

to an amplifier because the effect of the load impedance on the transducer can change its operating characteristics. For instance, applying a low impedance load can require more current to flow from the device output than it can deliver, causing the voltage to drop. (Figure 5-3 shows the voltage transferred as the ratio of load to source impedance increases.) Because matching impedances may be most critical when electromechanical devices are involved, some microphone preamplifiers provide adjustable input impedances to allow optimal impedance matching for different microphones. Such adjustment is not needed between purely electronic devices as long as the load is sufficiently larger than the source impedance.

The impedance matching we desire when transferring voltage differs from the more general case in which we wish to transfer total energy, as we do with power transfer. In the case of energy, the optimum transfer occurs when the impedances at both ends are equal. In addition, if impedances are imbalanced at a junction, some of the signal energy can be reflected instead of flowing to the destination. This effect becomes significant with short wavelength signals like digital audio signals and clocks. When cable lengths exceed about 10% of the wavelength of the signal, effects known as **transmission line effects** begin to occur. We think of wire as an instantaneous transmitter of electrical signals when dealing with audio frequencies; however, at very high frequencies (short wavelengths), the velocity of electron movement becomes a limiting factor. Generally, electrical signals in a cable flow at about 80% the speed of light. When very fast voltage changes occur, the time required to reach the far end of a long cable means that the voltage measured along the cable is no longer the same everywhere. A transmission line is said to exhibit a **characteristic impedance**, the complex ratio of the voltage to current along the wire. For coaxial cables, this falls in the range of 50–75 ohms, and for twisted-pair cables, around 100 ohms. When a transmission line is terminated with the characteristic impedance, no reflection occurs. When improperly terminated, energy reflects back at both ends of the line and causes interference with the signal.

There is a reason to not use larger resistances than necessary in a circuit. The amount of noise generated by the heat content of a resistance is proportional to the square root of the resistance:

$$v_n = \sqrt{kTRB} \qquad\qquad 7\text{-}1$$

where v_n = RMS noise voltage, k = Boltzmann's constant (1.38×10^{-23} joule/$^\circ$K), T = temperature ($^\circ$K), R = resistance (Ω), and B = bandwidth

($f_{max} - f_{min}$) in Hz. At room temperature, a 1 kΩ circuit has a minimum of about 0.3 μV of noise; a 1 MΩ circuit generates about 9 μV of noise. Although this may seem like a very small amount of noise, when multiplied by 60 dB (1000$\times$) of gain in a microphone preamp, it becomes considerable.

SHIELDING

Impedance matching alone is not adequate to guarantee unaltered signal transfer. When devices are far apart, the wiring connecting them may pick up unwanted signals by inductive and capacitive coupling from other sources. One method of reducing electromagnetic coupling is shielding the wiring. Shields are conductive sleeves placed around the signal wires and connected to ground. Any voltage created by the electric field is shorted to ground and kept away from the internal conductors that transmit the signal. The principle is that of the **Faraday cage**, a metal enclosure that is grounded to prevent electromagnetic signals from entering the interior of the cage. This setup does not prevent magnetically induced currents from affecting the internal conductors, however. Inductive coupling results when the signal conductors intercept a changing magnetic field. The distance between the conductors determines how many magnetic lines of force contribute to the current. By tightly twisting the conductors, the magnetic coupling is reduced. This design also helps reduce the contribution of capacitive coupling, as you will see.

In Figure 7-1, we see the equivalent circuit that represents an unbalanced connection. The signal is present on a single conductor that is shielded. The shield is part of the circuit and must be connected at both ends to complete a circuit. Unbalanced connections are simple and inexpensive. They work over short distances and are used frequently in consumer electronic devices. Their shortcomings quickly become evident as the size and complexity of installations increase.

THE BALANCED LINE

To further minimize contaminating signals, a differential transmission system known as a **balanced line** may be used (Figure 7-2). In a balanced line, signals are transmitted over two separate conductors with one carrying the signal and the other carrying an inverted version of the same signal. At the receiving end, the inverted signal is subtracted from the noninverted

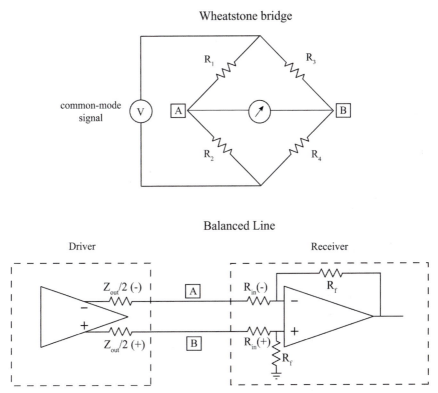

Figure 7-2 The Wheatstone bridge and the balanced line interface. The common-mode rejection ability of the balanced interface depends on the impedance balance between driver and receiver, which form a Wheatstone bridge. Points A and B show how the balanced interface is equivalent to a Wheatstone bridge.

signal, effectively doubling the amplitude of the recovered signal. In addition, any signal common to both received lines is canceled by the subtraction. Tightly twisting the wires causes electrical coupling to affect both wires equally, so the interference is equal on both conductors. This design is known as a **common-mode signal**. The common-mode cancelation is effective only if the impedances are tightly matched, or balanced. The **Wheatstone bridge** circuit helps explain why this is true.

The Wheatstone bridge is a circuit that can be used to determine the value of an unknown electronic component by matching it with known values. In Figure 7-2, bridge resistors R_1 and R_2 form a voltage divider, as do R_3 and R_4. R_1 and R_2 values are known, so the ratio of R_1 to R_2 is known. By measuring the voltage across the bridge, the ratio of R_1 to R_2 can be matched by R_3 and R_4. When there is no voltage measured, the bridge is in balance and the resistance ratios are identical. In our balanced line driver/receiver circuit, if the impedances are in balance, no common-mode voltage appears across the input of the receiver, which takes the place of the meter. The voltage V driving the bridge represents the common-mode voltage that would be produced in the connecting wires by induction or capacitive coupling. For the common-mode voltage to cancel, R_1/R_2 must exactly equal R_3/R_4.

Even tiny impedance mismatches can destroy the balance of the circuit and therefore its ability to cancel common–mode signals, a characteristic known as the **common–mode rejection ratio** (CMRR) of the amplifier. Transformers are the best way of providing the balance needed because their common-mode impedance is very high, making them less sensitive to imbalances in the source impedances. They are more expensive than op-amp differential amplifiers and less commonly used in modern audio equipment. Matching the resistors (R_{in} and R_f in Figure 7–2) used in an op-amp receiver to 0.1% can provide acceptable CMRR, and specialized integrated circuit line receivers now exist that can come close to the common-mode rejection possible with a transformer. The CMRR still ultimately depends on how well the output impedances are matched. Some manufacturers use a technique called **ground compensation** to create a pseudobalanced output by driving the positive output and tying the negative output to ground through a resistor intended to match the output impedance of the actively driven side. This setup rarely delivers the performance we get from a truly balanced output. It takes only a few ohms of imbalance to reduce the CMRR by 20 to 30 dB at an op–amp differential amplifier input.

GROUNDING

Recording studios are composed of numerous devices that must be interconnected, often over considerable distances. In addition to eliminating unwanted signal coupling through induction or capacitive coupling, the studio wiring must present the same ground potential to every device, because each device references its operation to the voltage present on its ground reference connection. If these differ between devices, their output signals will differ by the offset in their reference potentials. Should an AC signal appear on the ground line, that signal also appears in the output of any device sharing that ground reference. The longer the connecting wiring, the more difficult it becomes to guarantee that every device senses the same potential on its ground reference.

The most frequent contaminant of ground wiring is the power line frequency, 60 Hz or 50 Hz, depending on where you live. The slight hum often present in recordings usually results from grounding problems. The best way to guarantee proper ground references is the system known as the **star ground**. In the star ground system, each device has a separate connection to the ground point and no other connections to each other's grounds. This design is usually impossible to construct, in part because – for safety reasons – the power connection provided to all devices carries a separate ground line intended to protect in case of electrical faults in the equipment. This power line ground conducts fault currents away from the device and its users, but it also provides a separate ground circuit. Power line grounds are daisy chains, in which each outlet connects to the next in the circuit, rather than star grounds. Once there are two different points connected to a device ground input, a circuit is created that can conduct current and therefore produce a voltage difference between the devices by Ohm's law, discussed in Chapter 5. These ground circuits are called **ground loops**. The balanced connection has the advantage that the shield need be connected only at one end to provide electromagnetic protection, so the shield does not contribute to a ground loop. Unbalanced connections must connect the shield at both ends, making elimination of ground loops more difficult. In practice, detecting and eliminating ground loops can be a most frustrating activity.

The other critical characteristic of an amplifier is the gain, or amount of amplification, that the device can provide. Most amplifiers provide adjustment of the gain amount in order to handle signals of different amplitudes. How the gain adjustment is made can vary: the circuit gain may be adjustable or a fixed gain may be provided with a level adjustment before or after the fixed gain stage. How this is accomplished makes a difference: a fixed-gain amplifier is more easily overloaded if the gain adjustment is made after the gain stage instead of before. A further characteristic of note is whether the polarity of the amplified signal is the same as the input signal: some amplifiers, especially op-amp circuits, invert the signal, and others do not. Amplifiers can also affect the relative phase of signal components, especially when bandwidth is limited. This effect can result in timing differences between frequency components; the relative phase of each spectral component may not be preserved as the signal passes through the circuit. Although the human auditory system is not particularly sensitive to absolute phase, it is often easy to hear the effects of relative phase shifts when altered signals combine after being processed, especially with equalization. Components shifted in time and later mixed with the original signal create cancelations and reinforcements that can be audible even though the phase shift itself is not.

LINEAR AMPLIFIERS

These characteristics describe the gross level performance of amplifiers, but as we're interested in how amplifiers actually sound in practice, we must concern ourselves with some more detailed characteristics and how these contribute to the sonic performance of the amplifiers. We have so far assumed that any amplifier is completely distortion-free and alters nothing in the original signal. Such amplifiers do not exist in the real world. Real amplifiers deviate from the ideal in linearity, distortion, and noise, and each of these contribute to the sound. The significance of these limitations varies depending on the function of the amplifier: microphone preamps have different requirements from summing amplifiers or power amplifiers. Although all of these amplify, they are each optimized for their specific application.

Microphone amplifiers, usually called **preamplifiers**, are among the most critical of audio circuits. They must deliver high gains (as much as 70 dB), interface with a wide range of microphone types, and provide excellent common-mode rejection. Because any signal recorded through a microphone must also pass through a preamp, the sound we eventually record will contain any alterations generated by the preamp and its interaction with the microphone. Often, a specific preamp will sound different on a particular microphone than other preamps, and the preference between the two may also change from one recording job to another. Sometimes the difference is subtle and sometimes it is quite noticeable, so experimentation is frequently required.

Microphone loading by the preamp is most acute with dynamic microphones, in which the load interacts directly or through a transformer with the electromechanical element. If the preamp input impedance is too low, it will draw more current than the transducer is intended to produce, reducing the signal voltage. Ribbon microphones, with their low output levels, require high gains as well as proper impedance matching to produce the desired output. Moving coil dynamic microphones such as the Shure

SM57 are quite sensitive to the preamp load and can sound dramatically different into a simple solid-state preamp and into a transformer-coupled, well-designed preamp. Engineers are sometimes surprised how ordinary dynamic microphones can sound when used with a preamp that is optimally suited to that particular microphone. Because capacitor microphones have built-in amplifier circuits and produce higher output levels than dynamics, they are less sensitive to loading and do not require as much preamp gain. In fact, the high output level of capacitor microphones can cause overloads in preamplifier input circuitry and must sometimes be used with attenuators before the preamplifier input.

It is commonly believed that tube amplifiers are inherently different-sounding from solid-state amplifiers. The way amplifiers overload is different between tube and bijunction transistor devices. Tube and FET amplifiers are inherently current-limited and slowly reduce their output voltages as their outputs approach the limits of the power supply voltage. Transistor amplifiers are voltage-limited and quickly clip the output voltage when it exceeds the power supply voltage. However, when circuits are properly designed and constructed, the sonic differences are relatively small. The way the gain devices function does create some differences in noise and distortion characteristics, but both types can produce quiet and clean amplification or gritty distorted sound depending on how the circuits are implemented. Because both types of sound are sometimes desirable, we want to be able to select the type of amplifier that best suits the job we're trying to accomplish.

Voltage amplifiers can be classified in many ways: according to the gain element type, the circuit topology, and the function of the device (microphone preamp, power amp, etc.). After the choice of the active element is made, the circuit topology determines much of how the circuit will sound. We can classify amplifiers according to what portion of a sine wave signal the output device(s) conduct current. A transistor or tube running from a single voltage supply is called class A and its signal never swings through zero volts: the output device conducts current for the entire 360° of in input sine wave. The input signal is offset, or biased, so that the output swings around a voltage halfway between the power supply voltage and ground. Although this causes large power dissipation because current is always flowing, it produces a linear output because the voltage never crosses zero volts. A more efficient circuit uses complementary devices in a push–pull arrangement and requires a bipolar power supply so that the signal can swing negative as well as positive. Each output device conducts current only on half the 360° sine wave input and these amplifiers are called class B. Class B amplifiers can distort when the devices turn on and off at the zero crossing. Some amplifiers are intermediate between these types and are called class AB, with each output device conducting on between 180° and 360° of the input sine wave. These topologies trade off between zero-crossing distortion and high power consumption. Many of the most desirable amplifiers, regardless of function, are class A.

Another consideration related to the circuit topology is that of feedback. Many amplifier designs rely on negative feedback to stabilize the performance of the amplifier by subtracting the errors in the output from the input. Op-amp circuits offer a popular and simple example of the use of negative feedback, but feedback is not limited to op-amp circuits. By subtracting a portion of the output signal from the input, distortion of the amplifier can be improved. This approach can also degrade the transient performance of the amplifier if the feedback network introduces time lags between the output and input. Because tube circuits may be used with little or no feedback, they may not exhibit the negative effects of the feedback circuitry used in many solid-state designs.

In addition to the topology, the use of transformers can contribute to the sound of an amplifier, in both good and bad ways. The saturation of the iron core by the magnetic field imparts a characteristic sound to circuits using transformers. Well-designed transformers minimize low-frequency distortion and produce a wide bandwidth, but cheap, poorly designed transformers can degrade both of these factors. Good-quality transformers are expensive yet are desirable not only for their "sound" but also because they are useful in eliminating ground loops from amplifier input stages and providing excellent common-mode rejection. They also provide optimal loading for certain types of microphones, notably moving-coil and ribbon dynamic designs that are sensitive to loading effects. Because transformers are able to provide voltage gain, they can increase the overall gain of an amplifier without demanding more gain from the active circuit elements and may be used to match impedances between stages of amplification.

Another consideration, especially in input amplifier stages, is headroom: how much voltage the circuit can provide above that required by the signal's voltage swing. Amplifier circuits using vacuum tubes have power supplies producing hundreds of volts and therefore can provide more headroom than solid-state circuits, which are generally capable of providing around 30 volts or so. It is possible to use high-voltage transistors, but most audio amplifiers run on $+/-$ 15 to 18 volts if they use op-amps. Although this can be

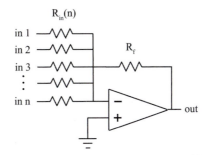

Figure 7-3 The op-amp summing amplifier.

sufficient, care must be exercised in the design to guarantee optimal signal levels. Headroom is also critical in summing amplifiers in which many signals are combined.

Summing amplifiers are the heart of mixers. They are used to add many signals together for the final output. Often dozens of separate signals must be combined. The summing amplifier must be able to handle a variable number of input signals without changing its performance. A popular circuit for summing is the inverting op–amp circuit of Figure 7-3. Each input is connected through a separate input resistor. As the number of inputs increases, the gain of the amplifier increases because the gain is the ratio of R_f to the total parallel resistance of the input resistors, which decreases as we add more inputs. There is a limit to the gain available, so there is also a limit to how many inputs may be summed on one op-amp. This fact sometimes limits the mixing performance of large mixers using op-amp summing amplifiers.

The choice of amplifiers can be made on many grounds, but the final decision is one of sound. We sometimes want the straight-wire-with-gain approach and sometimes prefer an amplifier with "character," one that alters the sound in a desirable way. By understanding the performance characteristics of the particular amplifiers we are using, we can obtain the behavior we desire.

NONLINEAR AMPLIFIERS

Most amplifiers are linear devices, but there are nonlinear amplifiers in common use: compressors and expanders. These amplifiers provide variable gain that can be controlled by the signal amplitude envelope. The idea is similar to what an engineer does with faders on a mixing board: raising the gain for quiet sounds and reducing it for loud ones. Although engineers become adept at making such changes, electronic circuits can be much faster and more accurate when properly designed and adjusted. There are many approaches to variable gain circuits and to the circuits that extract the envelope from a signal, each of which determines part of what the device sounds like and how it may be used. Compressors can be difficult to master without an understanding of how the particular device we are using accomplishes its work. By examining some popular compressors, we will explain why the different types sound different and how to decide which to use for a particular sound.

DYNAMIC RANGE PROCESSORS

The primary components of any dynamic range processing circuit are a variable gain element and a level sensing circuit. The level is determined by extracting the envelope of the signal, which is then used to represent the amplitude of the signal. This determination can be accomplished by rectification and filtering, creating a measure corresponding to the average, or it can be computed by a root-mean-squared circuit that more closely resembles the loudness we perceive. Neither the average nor the root-mean-squared circuits will reflect the maximum signal voltage, so peak-detecting circuits are sometimes used as well. The different envelope-detecting approaches used in compressors and expanders account for some of the difference in how various devices sound. There are also many approaches to the variable gain element, ranging from purely electronic elements, including FETs, vacuum tubes, and voltage-controlled amplifier (VCA) integrated circuits to electro-optical elements that use light modulated by the signal amplitude envelope falling on a light-dependent resistor (LDR) or photo-transistor. Each has advantages and disadvantages as well as a characteristic sound.

Optical elements tend to be slower in their action than electronic gain elements. Typical optical compressors like the LA-2A sound smooth on vocals and bass guitar. The optical gain element consists of a light source driven by the signal and a LDR or phototransistor whose resistance depends on the amount of light that falls on it. Faster compressors using FET gain elements like the 1176 are more aggressive on transients like drum sounds. Compressors using a variable-mu gain element, a specific kind of vacuum tube whose gain is changed by using the control voltage from the detector to change the bias, also produce a characteristic sound. There are many variations possible when combining detector circuit types with gain element types, allowing a wide range of compressors and limiters that all sound different.

In addition to the particular circuitry used for signal level determination and gain control, there are two basic topologies for routing the signal to the measurement circuitry: feed-forward and feedback. When the input to the gain processing device is taken as the signal to be measured, it is a feed-forward circuit, when the output signal is used it is a feedback circuit (see Figure 7-4). There is an essential difference in the behavior of these two topologies because one measures an unaltered signal and the other measures an already compressed or expanded signal. This difference also results in characteristic differences in the sound of the devices.

Compression and limiting are two of the more powerful processing techniques available for music mixing and production. Careful use of dynamic range processing can result in intense yet natural-sounding mixes. It often accounts for much of the difference between amateur and professional productions. Although the two techniques are essentially the same conceptually, they have different sonic characteristics. Compression using low to moderate ratios requires lowering the threshold in order to affect the dynamics audibly, which changes the amplitude relationships of the low and mid-level sounds. Limiting generally requires a higher threshold, which affects only the peaks of the signal and does not alter the relationship of low and mid-level sounds the way compression does. The result is a louder sound with the original dynamic relationships for the lower amplitude sounds.

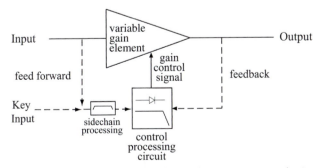

Figure 7-4 Feedback and feed-forward compressor topologies.

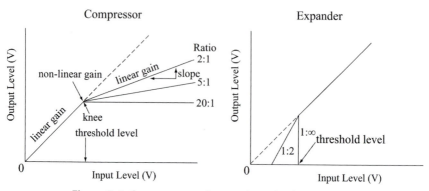

Figure 7-5 Compressor and expander gain characteristics.

Compressors function by comparing the measured signal level against a reference level we set, that is, the threshold. If the signal exceeds this level, the gain is reduced. When the signal drops below the threshold, the gain is restored. Figure 7-5 shows the gain characteristics below and above the threshold. The gain of the compressor is linear below and above the threshold; it is nonlinear only at the threshold where the gain changes. The sharpness of this transition is described by the characteristic "knee"; a sharp transition is called a hard knee and a smooth transition called a soft knee. The above-threshold gain is the slope of the line, the ratio of output to input level: the higher the ratio, the more compression. At high ratios, those greater than 5:1, the effect is known as **limiting** because the output level is effectively unchanged with increasing input level – the output becomes "limited" to a maximum value set by the threshold.

The attack and release times determine how quickly the gain is changed when the signal exceeds the threshold and when it drops below the threshold, respectively (Figure 7-6). These timings dramatically change the sound we hear: fast attacks remove transients from the signal, and longer attack times permit more of the onset of sounds to be heard while the more continuous elements of the sound are reduced. At extremely short attack and release times, each cycle of low frequency signals may trigger and release the compressor, producing distortion. With longer attack times, the compressor does not fully reduce the gain immediately and the signal overshoots the final level. Adjusting the attack and release times gives us control of the dynamics of the signal as the amplitude moves up and down. If the attack and release are set very short, there is audible pumping as the gain changes abruptly and frequently. One strategy for natural-sounding

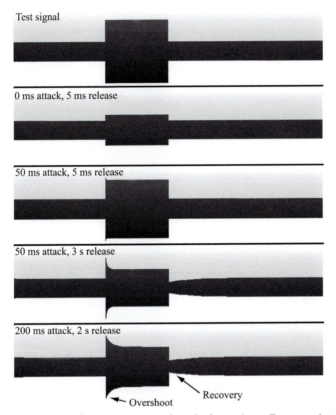

Test signal

0 ms attack, 5 ms release

50 ms attack, 5 ms release

50 ms attack, 3 s release

200 ms attack, 2 s release

Overshoot   Recovery

Figure 7-6 The effects of compressor attack and release times. Top trace is test signal of a 1 kHz – 13dBv baseline, 10 dB step. Compression ratio is 4:1 and threshold is –6.5 dBv.

compression is to take advantage of the attack and release times of the human acoustic reflex, which has attack times of about 40 milliseconds and release times of about 135 milliseconds. Short attack times can be helpful in preventing transients from overloading the audio system when limiting is desired.

The inverse process to compression is expansion (Figure 7-5, right); as the gain drops below the threshold, the gain is reduced. Above the threshold, the gain is linear. The extreme case of expansion is **gating**, in which the sound is turned off when the level drops below the threshold. This technique is often used to eliminate low-level noise from a sound. Expansion can also be used to adjust the decay of percussive sounds by reducing the decay time using the expander's release time controls. Both compression and expansion allow the engineer to alter the amplitude envelope of individual sounds as well as to modify the dynamics of an entire mix. Manipulating dynamic range is also possible through less controlled means, as by intentionally overloading amplifiers or analog-to-digital converters to produce a louder-sounding signal at the cost of increased distortion. Several stages of compression may be combined, either in series or in parallel. Compressed signals may be combined with the uncompressed version to restore some dynamics removed by the compression. The combination of these dynamic range processing techniques contributes much to the sound of popular music recordings.

EQUALIZERS

In addition to the manipulations of amplitude dynamics, the spectral content of a recorded sound may be altered using the techniques of equalization. Equalizers are amplifiers that allow bands of frequencies to be separated and altered individually. The equalizer band is a band-pass amplifier that changes the gain for a limited band of frequencies. There are many ways to implement such devices: the bands may be fixed overlapping band-pass amplifiers with gain adjustment (graphic equalizers); individual amplifiers in which the parameters of bandwidth, center frequency, and amplitude are individually adjustable (parametric equalizers); or shelving filters that set a corner frequency above or below which the amplitude may be adjusted. Equalizers may also be semiparametric, allowing the center frequency and gain to be altered with a fixed bandwidth. Simple high-pass and low-pass filters are also available. Parametric (or semiparametric), high-pass/low-pass, and shelving types are often combined, especially on mixing board channels.

Figure 7-7 shows the parameters of a fully parametric equalizer amplifier. The center frequency (f_0) is adjusted to set the frequency placement of the filter, "Q" or the quality factor determines the width of the frequency band affected, and the gain is increased or decreased with the gain control. Note that here zero gain means no amplitude change, not no output. Some filters are capable of resonance, actually oscillating at a particular frequency when driven by a signal at that frequency. These filters are used often in synthesizers to create dramatic effects but may also be employed in mixing, especially when using digital filter plug-ins. Generally, such oscillations are not desirable in amplifiers.

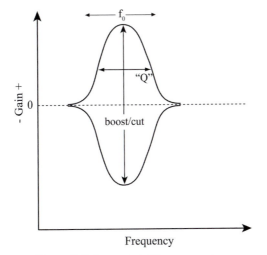

Figure 7-7 Parametric filter parameters.

Figure 7-8 shows the effect of filter Q on the phase response for a two-pole low-pass filter. Each pole contributes 6 dB/octave slope to the overall filter shape. As the Q is broadened, the frequency range affected by phase shifts broadens. Multiple-pole filter designs can be quite complex in their phase and magnitude behavior as each pole interacts with the others. These filters may be optimized for flat frequency response within the pass band at the expense of phase behavior or for minimal phase shift at the expense of in-band ripple in the frequency response. The diversity of filter designs

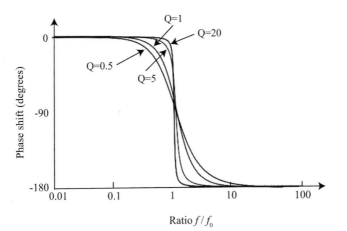

Figure 7-8 Phase shift as a function of Q for two-pole low-pass filter.

available provides a range of different sounding options for filters of similar functionality. Although we are relatively insensitive to absolute phase, we can hear interference if phase-shifted components are added back to the original signal.

MIXING CONSOLES

The heart of most studios is the mixing console. These are large collections of individual amplifiers that allow many signals to be routed, processed, and mixed together. There is great variety in the design and construction of mixing boards, but they are all collections of amplifier circuits that may be interconnected to create the particular functionality needed to mix and process individually recorded tracks into a stereo or surround output. The complexity can be daunting, but realizing that each channel strip is a functional device made up of several equalizer amplifiers and perhaps a dynamic range processor will help simplify the system for you. The individual channels may be connected to a bus, a **summing amplifier** that adds together multiple inputs. The bus also often allows processing of its combined signal before it is routed to an output. Subbuses are designed to allow a subset of selected individual tracks to be mixed together for processing before routing to a full mix with an individual level control. Effects sends are dedicated buses for combining signals to allow group processing by internal or external devices such as reverbs, delays, or other signal processors. Mixers generally have from 2 to 24 buses, with 8-bus designs common for recording. Each bus may have an insert connection with which an external signal processor may be patched, or temporarily connected, to allow additional processing not provided by the mixer itself.

Many mixers, particularly recent digital models, also have automation capabilities. This feature allows the mixer to remember the individual control adjustments over time as they are changed and restore them later, allowing for some tricky changes that would be difficult without some mechanical help. With only ten fingers, an unaided engineer cannot manipulate more than a few controls at a time; automation can simultaneously move as many as necessary. It also allows later return to a mix so that small changes may be made without redoing the entire mix. This ability is a convenience, but it does not change the basic functionality of a mixer, which is determined by the number of channel strips and the internal wiring connections provided. There are two main designs in common use: the split console and the inline console. The split console consists of two separate mixers: one for handling inputs from microphone and line inputs to feed the recorder and another to mix the recorded tracks for monitoring. On an inline console, each channel strip handles both the incoming signal and the return from the recorder that may be separately monitored. Although the tape return section is physically part of the inline channel strip, it is functionally separate, but it may sometimes use some of the strip's equalizer sections. The tape monitoring section of an inline console can be considered a separate mixer, like the split console monitoring section, incorporated into the main channel strips. Inline designs are less expensive to build and are the most popular type for small to mid-size mixing consoles.

Although the size and complexity of mixers may be significant, the basic concept is quite simple and is much the same for all mixing consoles. This knowledge should relieve at least some of the anxiety people experience when first using a massive unfamiliar mixer. Just remember that it is a collection of simple amplifiers with defined routing possibilities between functional blocks – not the control panel from an extraterrestrial flying vehicle.

SUGGESTED READING

Ballou, G. M. (2008). *Handbook for Sound Engineers* (4th ed.). Focal Press. ISBN: 0-240-80969-6. This volume should be in every serious engineer's library.

Analog Recorders

Contents

Sound recording has taken many forms as technological developments have created new and improved ways of storing information. The first sound recording used smoked paper and a stylus to record sound, but no way of playing back the sound existed until recently. The earliest practical attempts to capture and reproduce sound used mechanical methods, such as Edison's machines, which connected a large horn to a tiny stylus that traced grooves in a rotating cylinder. Optical systems have long been used for film sound, but analog optical systems are complicated to record and tend to be noisy. The most flexible system of sound recording has been the magnetic approach, in which the sound is converted first to an electrical analog and then to a proportional magnetic flux that leaves a pattern of magnetization on a medium that can retain the signal for a long time.

The earliest magnetic recorders used steel wire or ribbon, an obvious approach based on the understanding of iron's ability to be magnetized. However, steel wire was inconvenient, not to mention dangerous, so the practical magnetic recorder would have to wait for the development of better magnetic materials. The breakthrough that opened the way to modern recorders was the use of magnetic oxides glued to flexible tape. This development allowed the creation of reliable recorders with easily inter-changeable media and started the tape recorder revolution. The hard drives we use today are directly descended from magnetic tape and, although the encoding is different, operate under the same physical constraints as analog tape recorders.

The Science of Sound Recording
ISBN 978-0-240-82154-2

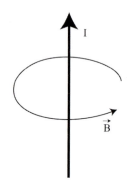

Figure 8-1 Current I produces magnetic field $\vec{B}$

MAGNETISM

An electric current always produces a proportional magnetic field, a basic property of electromagnetism. Current flowing in a straight wire generates a magnetic field that curves around the length of the wire and spreads out in space, diminishing in amplitude with the square of the distance from the wire. The magnetic field orients in what's known as the "right-hand rule": if the current flows in the direction of the right thumb, the magnetic field is in the direction of the curled fingers. If the wire is instead wrapped around a piece of iron, a magnetic field will be generated within the iron, creating an electromagnet. Iron has a low resistance to the magnetic flux (the magnetic equivalent of electric current), which is called **reluctance** (the magnetic equivalent of resistance). The type of magnetization produced in iron is known as **ferromagnetism**. This type of magnetization can persist after the applied current or external magnetic field is withdrawn, making it ideal for storing magnetic patterns. Other types of magnetization (paramagnetism and diamagnetism) occur but do not persist after the applied force is removed and are therefore not of interest in the pursuit of making recordings.

If the metal core of an electromagnet is shaped into a ring with a small gap, the magnetic flux will flow in the iron until it reaches the gap, at which point it will flow into space in all directions. If we then introduce a magnetic medium in close proximity to the gap, the magnetic lines of force will tend to flow through the magnetic material because it has less reluctance than the air (see Figure 8-2). As the magnetic material is moved past the head gap, there will be a pattern of magnetization written on the medium as the magnetic domains – functionally, tiny bar magnets – align

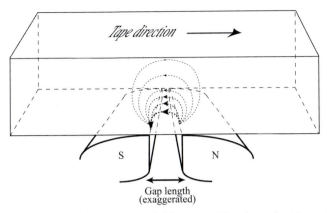

Figure 8-2 Part of the magnetic field at record head gap (not to scale).

their fields with the applied field. Once the applied magnetic field strength is reduced to a level insufficient to further change the magnetization of the medium, the magnetization pattern left on the medium will be retained. The actual recording occurs at a distance from the record head gap because the applied field spreads out in space and the point at which the field strength drops below the threshold necessary to further alter the polarization of the domains occurs away from the actual gap. The system requires the medium to be moved past the writing head at a known and constant rate. The geometry of the head, its gap, and the magnetic medium will determine how accurately the signal is written. It is important to understand how magnetic fields behave as this is fundamental to how a signal can be transferred to magnetic media.

Magnetic media are made of tiny particles of magnetic materials glued to a substrate. The particles do not move physically but rather adjust their magnetic polarities to align with an applied field. The basic structural element of magnetization is the domain, which is conceptually a tiny magnet with a north and a south pole, referred to as a **dipole**. Domains are areas of a magnetic material in which the unpaired outer electrons' spins interact to produce aligned magnetic dipoles. Domains do not correspond exactly with particle size or shape, and many domains may be contained within a single particle. Each of these domains can be aligned in one of two opposite directions (north–south or south–north) by an applied magnetic field. Only those domains that happen to physically align with the applied field will be susceptible to changing their polarizations. If there are enough of these domains attached to the medium in random orientations, patterns of retained magnetization can be created by moving the medium past the applied field while it is changing in proportion to the electrical signal. As the applied magnetic field decreases, with either distance or decreasing signal amplitude, the magnetization drops below the strength needed to further alter the domains polarizations and the existing signal is stored.

The field strength required to alter the polarization of a medium is called the **coercivity** of the medium, a measure of how hard it would be to record or erase the magnetic signal. Some magnetic materials are easy to magnetize and lose their magnetization easily. These are called **soft** magnetic materials and are used in transformers and sensors. Other materials are more difficult to magnetize, retain their magnetization better, and are known as **hard** magnetic materials. These are employed in motors, generators, and actuators like solenoids. In between are the materials used for magnetic recording, for which intermediate coercivities allow recording and erasure at field strengths produced by magnetic heads while retaining the recorded magnetization.

Although we are most familiar with magnets made of iron and alloys with other metals, the atoms of certain other elements are capable of ferromagnetization: nickel, cobalt, chromium, manganese, gadolinium, and dysprosium. Atoms of these elements exhibit the quantum electron spin effect known as **exchange coupling**, which allows the magnetic dipole moments of adjacent atoms to align into a magnetic field. Because the atoms of these elements produce the magnetic field, salts of these elements are also capable of magnetization. These salts, mostly oxides, are used as magnetic materials in magnetic tape and discs. Newer tape formulations used for video and digital audio use metal particles and evaporated metal coatings to achieve higher coercivities.

THE PHYSICS OF MAGNETIC RECORDING

The fact that there exists a minimum magnetic field strength required to begin altering the polarity of the domains leads to one of the limitations of magnetic recording. Because there is no effect when the signal strength falls below the threshold for magnetizing the material, low-level signals are not recorded. At very high field strengths, all of the susceptible domains are polarized and no further signal amplitude can be recorded. This polarization leads to a compression effect that can be exploited in analog recording, but it is a nonlinear distortion of the signal. Once the maximum level has been recorded on the medium, reversing the field does not immediately cause a reversal of the magnetization of the medium until the applied field exceeds the threshold amplitude required to change the magnetic polarization previously written to the medium. This phenomenon is known as **hysteresis** and causes yet another nonlinear form of distortion.

In Figure 8-3, H is the applied magnetic field (in oersteds, Oe = 79.6 A/m) and M (Oe) is the retained or remanent field. As the applied field strength grows in the positive direction, it begins to change the magnetic particles' magnetic field alignments, increasing until no domains remain to be magnetized. As the applied field strength decreases, the magnetization remains until the reversing applied field exceeds the threshold for reversing the retained magnetization. Though some of the curves have linear regions, the extreme low- and high-level signals will be distorted in such a system. In the case of digital recording, for which we need to record only two discrete levels or polarization directions, the nonlinearity is not so important, but for analog recording, it is a serious problem.

Bias is a term for offset, a DC voltage or current added to change the signal baseline, often to center it in the range of a circuit voltage. Early magnetic recorders used **DC bias** to offset the signal along the initial rising curve and utilized only the small linear segment. Because the magnetization always produced a field with the same direction, it resulted in the eventual

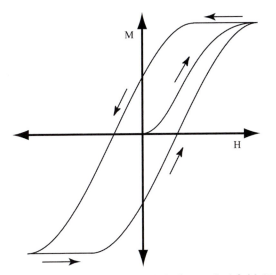

Figure 8-3 Magnetic hysteresis (M-H) curves. H is the applied field; M is the retained magnetic field.

magnetization of the heads themselves. It also resulted in noisy recordings because so little of the magnetic hysteresis (M-H) curve could be used. The discovery of **AC bias**, in which a large-amplitude high-frequency sinusoidal signal is mixed linearly with the audio signal, made the entire range of the curves accessible and allowed the development of modern high-fidelity magnetic recording. The bias signal frequency is five to ten times higher than the highest audio frequency, and it is recorded along with the signal, improving the linearity of the audio signal reproduced. Many explanations have been offered about how AC bias acts to linearize the hysteresis curve because the discovery was reportedly an accident, albeit one that opened the door to the modern era of sound recording. To understand what AC bias does to eliminate the nonlinearities of the hysteresis curve, we will examine the different ways of explaining the effect.

If we try to record a sine wave signal directly with no bias, we find the region of the signal around the zero field strength level is not able to cause any magnetic particles to reorient their polarization, and therefore no signal is recorded (Figure 8-4). As the signal increases, the magnetic field applied by the head starts to exceed the threshold for magnetizing the medium and a signal is recorded. As the amplitude increases further, more domains are affected and more magnetization is left on the medium. As the signal increases further, it eventually magnetizes all of the susceptible domains and

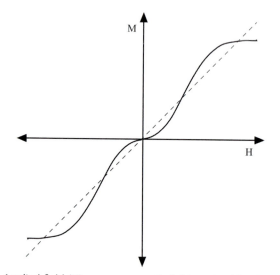

Figure 8-4 Applied field (H) versus magnetic field retained by the medium (M).

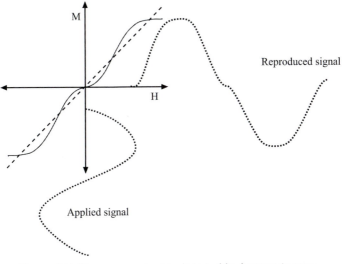

Figure 8-5 Sine wave output is distorted by hysteresis curve.

no further signal can be recorded. The result is a sine wave with a distorted area around zero crossing and a flattened top (Figure 8-5).

If we add a large-amplitude, high frequency sinusoidal signal to the audio signal, the composite signal can be fed to the record head and used to magnetize the medium. Because the two signals are simply added together, they can later be separated with a simple filter. The resulting audio signal is played back with greatly reduced distortion. Though the bias signal is itself distorted at the zero crossing, the envelope of the bias signal plus the audio signal is not affected. Any bias signal reproduced is removed by filtering, and the remaining audio signal is recovered without zero crossing distortion.

AC bias has been described as a reduction of the inertia involved in aligning the domains with the applied field – a sort of "shaking up" of the magnetic domains so that it is easier to affect their magnetic polarization. Because the domains do not physically move, this explanation is a bit confusing, although it does address the issue of low field strengths failing to alter the magnetization of the medium. What bias current does is increase the average level of applied field, enabling the magnetization of low-level signals that would otherwise be too small to change the polarizations of any domains. The effect of AC bias is referred to as anhysteretic magnetization. In the reproduction process, the high-frequency bias signal is filtered out by a notch filter, the **bias trap**, without altering the audio signal's frequencies, resulting in a linearized response to the low-level signal. Care must also be taken to prevent the bias signal from bleeding into other parts of the record circuitry; bias trap filters are therefore used to keep the bias signal out of the record electronics.

Adjustment of the amount of bias added to the record signal gives the engineer the ability to select a balance between noise performance, distortion, and bandwidth. By increasing the amount of bias current from a low level while recording a test sine wave signal, a peak in the amplitude of the retained signal can be determined by watching the playback meter rise, peak, and begin to fall. The peak indicates the maximum sensitivity of the tape at the frequency of the test signal. Increasing the bias beyond the peak level will decrease the amount of high frequencies stored while also decreasing distortion and noise. The optimum amount of bias used varies with the type of tape used. Each tape formulation has specific bias settings for minimum distortion, highest output sensitivity, and minimum noise. The optimum bias setting for one characteristic is rarely optimal for others. Tape manufacturers will recommend a specific amount of "overbias" be used to provide the claimed performance of the tape, but such a determination is open to the interpretation of the user, who may choose a different balance of noise and distortion from that recommended. The creative use of bias adjustment is discussed later in the section dealing with recorder calibration procedures.

THE INTERACTION OF HEAD AND TAPE

The magnetic field applied by the record head interacts with the tape in a complex way. Because the tape has a finite thickness, magnetic particles at different depths in the tape experience different applied field strengths. This effect is wavelength dependent and therefore varies with frequency. Because the signal is actually recorded where the applied field decreases to the threshold for domain magnetization, it occurs at a distance from the head gap itself – this distance is the so-called "trailing edge." This distance is also wavelength dependent. The transfer of magnetization from the head to the tape is critically dependent on the distance from the gap; thus, any separation of the tape from the head results in a diminished retained or reproduced signal, a phenomenon called a **dropout**. This effect is similar to the reduction of retained magnetization with increasing depth into the magnetic medium. The geometry of the tape and head determine much of the efficiency of the recording process. The same is true for the reproduction process, although the demands are different.

The equation relating the reproduced voltage (e) and the flux seen by the play head is:

$$e = -N\frac{d\Phi}{dt} = -Nv\frac{d\Phi}{dx} \qquad\qquad 8\text{-}1$$

where:

Φ = average gap flux

t = time

v = velocity

x = position

N = number of turns of wire in the head

This equation shows that the output voltage is a function of the rate of change of flux on the tape $d\Phi/dt$, making this a **differentiating process**. This process is similar to the behavior of an inductor, in which the voltage drop is proportional to the rate of change of the current. Here, the magnetic flux creates a current in the windings of the head by induction. Because the rate of flux change increases with increasing frequency (decreasing wavelength) of the recorded signal, the output voltage also increases with increasing frequency. (It will also increase if the tape is pulled past the head faster.) This results in a 6 dB/octave rise in output voltage. The differentiation created by the physics of the reproduce process requires that the reproduced signal be equalized at −6 dB/octave in order to restore the original signal. Other minor factors alter the frequency response of the process and require further equalization. Different "standard" equalization curves exist; tape and recorder manufacturers generally specify which they deem best suited to their products.

As the tape moves past the record head, a magnetic signal is transferred to the tape. The wavelength of the signal depends on the frequency and the tape speed. For a tape speed of 38 cm/s (15 inches per second (in/s)), a 20 Hz sine wave is 1.9 cm long; a 20 kHz sine wave is 19 μm in length, about the thickness of the tape coating and about the length of the head gap. The wavelengths change with the tape speed, so at 76 cm/s (30 in/s), the wavelengths double. We will see how the signal wavelength relative to the head gap influences the reproduction process.

The reproduce gap generates an output voltage that depends on the change in average flux sensed by the head gap. So when the wavelength of the signal is exactly the length of the gap, the average flux is one complete sinusoid, generating an average flux of zero. Therefore, a signal at that frequency will produce no output: the gap length of the reproduce head determines the maximum frequency signal that can be reproduced. In the case of the record head, the applied field strength causes the signal to be recorded – and the larger the record gap, the more field that can be generated and the more signal that can be recorded. There is a basic difference between the record and play functions, so the gap length plays a different role in the two cases. The record gap should be large enough to create a magnetic field deep into the tape magnetic coating. The erase head needs enough field strength to reach any areas of the tape left magnetized by the record process. The reproduce gap should be large enough to capture the most flux change but small enough to allow short wavelengths to be reproduced. Obviously, there is a trade-off inherent in the playback head gap dimensions.

An equation approximating the output voltage [$e(x)$] from a reproduce head is:

$$e(x) = -\mu_0 VwM_0(H_g g/i)k\delta[e^{-kd}][1 - e^{-k\delta}/k\delta]$$
$$\times [\sin(kg/2)/(kg/2)]\cos(kx) \qquad\qquad 8\text{-}2$$

where:

x	= longitudinal position (Vt [velocity x time])
$e(x)$	= voltage output from longitudinal magnetic recording
μ_0	= magnetic permeability of a vacuum
V	= tape-to-head velocity
w	= track width
M_0	= peak value of sine-wave magnetization
H_g	= deep gap field
g	= gap length
i	= current in head coil
k	= wave number (= $2\pi/\lambda$) where λ = wavelength
δ	= thickness of the magnetic medium
d	= distance from tape to head

As is clear from the equation, many physical characteristics of the playback head and tape affect the recovered electrical signal. In the following equations, the terms in square brackets are loss terms relating to specific physical relationships that act to reduce the output voltage at the reproduce head.

Spacing loss is a decrease in output signal due to distance from tape to head:

$$L_d(dB) = 20\log(e^{-kd}) \qquad\qquad 8\text{-}3$$

Spacing loss increases exponentially as the distance relative to the wavelength of the signal increases; thus high frequencies are more susceptible to dropouts. (Spacing loss can be expressed as 54.6 dB/wavelength distance, meaning that almost 60 dB of dropout is produced if the tape is separated from the head by the wavelength of the signal frequency of interest!) Preventing high-frequency dropouts caused by dirt particles requires that heads and transports be kept clean, as even microscopic particles cause significant dropouts.

Gap loss is a decrease in output signal due to the length of the gap:

$$L_g(dB) = 20\log\frac{\sin(kg/2)}{(kg/2)} \qquad\qquad 8\text{-}4$$

Gap loss reflects the fact that the reproduce head responds to the average flux in the gap; therefore, when the wavelength of the signal just equals the gap length, the average flux from one sine cycle is zero and there is no signal produced. The effective gap length is 14% larger than the actual gap length due to fringing of the magnetic field and other factors. Though gap loss is often assumed to be the dominant limitation to high-frequency reproduction, reproduce gaps of lengths commonly used in professional recorders do not produce significant loss, even at frequencies well above 20 kHz. For example, a recorder with a playback gap length of 2.5 μm produces a first null at over 150 kHz at 38 cm/s (15 in/s) and a −3 dB point of 61 kHz. Reproduce gaps range from 1.5 to 6 μm.

Where gap losses may become important is in the case of synchronous overdubbing, when the record head is used to play back tracks already recorded on other tracks. Because record head gaps are longer to generate large record fields, the lengths may significantly degrade their playback response due to gap losses. Record gaps range from 4 to 13 μm in modern machines. For a record gap of about 13 μm, the recorder will have a first null at about 30 kHz, well below the first null frequency of the reproduce heads. Erase heads use longer gaps to insure erasing signals the full depth of the magnetic coating, with lengths ranging up to 100 μm. Erase heads may also use multiple gaps to increase the efficiency of the erasure.

At low frequencies, the gap senses magnetic fields that originate at a distance from the gap, resulting in a series of peaks and dips in the frequency response known as contour effect or **head bump**. The magnetic field generated by the tape, especially at low frequencies, enters the heads from areas not directly over the gap in a phenomenon known as **fringing**.

As the frequencies in the fringing vary, they cancel and reinforce the signal frequencies, thereby producing the peaks and dips. The contour effect is additive as multiple generations of analog recording and playback are performed, so a final stereo mix may show several decibels of low-frequency ripple. The contour effect has been reduced by careful head design in later generations of analog recorders, but it is also responsible for some of the sound of earlier machines that were often used to record rock music.

Thickness loss is a decrease in output signal due to the thickness of the magnetic medium:

$$L_\delta(dB) = 20 \log \left[\frac{1 - e^{-k\delta}}{k\delta} \right]$$

8-5

Thickness loss (~16 dB/wavelength distance), though not a universally accepted designation, does indicate that the thickness of the magnetic layer is real and that it affects the reproduced signal to a measurable degree. As the field penetrates further into the magnetic coating, less magnetization is retained by the tape because the applied field diminishes with the square of the distance from the gap. In order to determine the overall playback signal amplitude, we must integrate the magnetization through the depth of the magnetic coating. This effect differs from spacing loss because here the nearest region of the coating is still in contact with the gap, while deeper layers are removed by increasing distance. The magnetic coating thickness for Ampex 456, a tape popular for recording rock music, is about 14 μm, producing about 13 dB of loss for a 19 μm wavelength (20 kHz @ 38 cm/s) sine wave. For high audio frequencies, the thickness of the magnetic coating is the dominant cause of signal loss – another of the reasons that equalizing the recording process is necessary.

There are many potential sources of distortion in the process of magnetic recording. Changes in both amplitude and frequency may occur that alter the original signal by the time it is played back. These both fall into the category of modulation and have been addressed frequently in the development of the tape recorder. Amplitude modulation (AM) results from changes in amplitude caused by dropouts or uneven oxide distribution on tape, for example. Frequency modulation (FM) results when the frequency of the signal is changed by tape speed variations caused by poor capstan motor regulation or scraping of the tape against guides or the heads themselves. These forms of modulation cause components to appear in the signal that were not originally present.

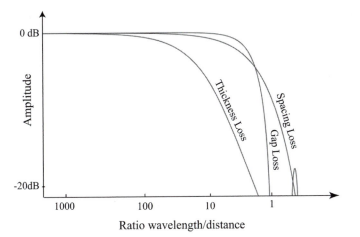

Figure 8-6 Amplitude losses in reproduced signal as a function of the ratio of distance from gap or the gap-length-to-signal wavelength. Thickness loss is dominant at high frequencies.

SOURCES OF NOISE

The noise performance of tape recording is dependent on the magnetic material used, the head design, and the bias and equalization settings chosen. There are several sources of noise, some of which relate to the physical construction of the medium and some of which relate to the heads and their interaction with the tape. The application of a DC signal to a recording head results in an increase in playback noise of up to 10 dB over the level on blank unrecorded tape. This phenomenon is related to the imperfect distribution of magnetic particle orientation and size. A perfect magnetic medium would have a random distribution of infinitesimal magnetic particles. Though this perfection is impossible in real-world manufacturing, the minimization of noise is one of the goals of continued research in materials science. Because AC signals can be considered as short sections of DC, the recording of AC signals also generates a noise known as **modulation noise**. This noise is clearly audible when recording pure sine waves (calibration tones, for example) as the sine tone fails to fully mask the sideband frequencies generated. The modulation noise is largely masked by complex musical signals. Recording and playing back a 10 Hz sine wave will clearly reveal the contribution of modulation noise because only the noise will be audible.

The physical smoothness of the tape surface determines the amount of **asperity noise**, which is generated by an uneven tape surface changing the head-to-tape contact. Minute imperfections in the medium surface cause the head-to-tape spacing to change and increase the friction of the contact. The nonuniform distribution of magnetic particles on the tape results in a form of noise resulting from flux lines caused by differences in magnetization along the tape even when the applied field is constant. Both of these noise sources can be minimized by reducing the size of the magnetic particles and by reducing the variation in particle size and distribution.

The record head also contributes to noise, as the domains within the head itself react to the imposed magnetic field and restrictions that limit their ability to smoothly align their magnetic polarities, creating what is known as **Barkhausen noise**. As each domain aligns and joins those already aligned, noise bursts result. By reducing the size of the domains, this noise can be minimized. A further source of noise in record heads results from magnetostriction, mechanical stress changes in the head material due to the magnetic field force on the metal. (This phenomenon also accounts for the hum heard in power transformers.) Some noise is also contributed by the electronic circuitry, particularly the preamplifier required to boost the voltage recovered by the playback head, although this noise is generally significantly lower than the noise from the magnetic medium itself.

NOISE REDUCTION SYSTEMS

Due to the noise inherent in analog recorders, add-on systems have been devised to reduce the noise level. Some such systems use dynamic low-pass filters to separate signal from noise, but most noise reduction devices use compression to reduce the input signal's dynamic range and complementary expansion on playback to restore the original dynamics. Though these systems work on similar principles, each approaches the problem slightly differently and with varying levels of complexity. Popular music recording often avoids using noise reduction by recording high amplitudes to tape, maximizing the inherent signal-to-noise ratio. Enough engineers prefer the lower noise to encourage the development of very complex multiband noise reduction systems that make use of auditory masking curves to maximize the noise reduction while minimizing audible alteration of the signal. Because each track added to a mix contributes 3 dB of wideband noise,

noise reduction is of more importance to multitrack recorders than simpler stereo recorders.

dbx type I noise reduction uses a single compressor/expander for each channel with a fixed preemphasis/deemphasis filter system. By boosting the high frequencies before recording the signal, the system is able to cut high frequencies a complementary amount on playback, reducing high-frequency noise added during the recording process at the same time. This effect relies on the facts that high-frequency components of the signal are usually of low amplitude and that the preemphasis will not cause clipping. Dolby A uses four filter bands, each with a separate compressor and expander. Dolby SR uses both fixed and adaptive filters to split the spectrum into ten separately processed frequency bands. The issue of recorder calibration becomes important with any of these devices, as they rely on system linearity to properly restore the original dynamic range of the signal. Calibration tones should be recorded at the beginning of tapes that include with noise reduction to allow proper frequency response adjustments to be made later. Improper adjustment of recorder frequency response will result in artifacts created by the noise reduction systems, usually audible gain changes that create a "pumping" effect.

TAPE RECORDERS

Figure 8-7 shows a modern multitrack tape recorder. The recorder is capable of recording 24 separate tracks. Extensive electrical and mechanical control systems contribute to the machine's ability to maintain constant tape

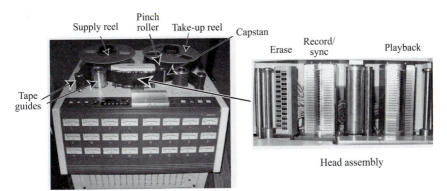

Figure 8-7 A 2-inch, 24-track tape recorder and its heads. The transport must spin a 10-inch reel of tape weighing 10 pounds while maintaining constant tension and speed. Constant tape alignment is aided by tape guides.

speed with a ten-pound reel of tape, making use of reel motors to maintain hold-back tension and a separate motor driving the capstan with a pinch roller to control the tape speed. The capstan rotation rate is measured and maintained within tight limits by servo feedback control. The tape path includes guides to maintain the tape alignment and filter out mechanical vibrations as the tape is pulled over the heads. Speed variations called **wow and flutter** (low- and higher-speed variations, respectively) cause frequency modulation of the signal if not controlled. The head assembly consists of three heads: an erase head, a record/monitor head, and a reproduce head. The heads must be carefully aligned with each other so that signals may be recorded, played, and erased properly. Each track is essentially a separate recorder with its own electronic adjustments, and full electronic alignment can take some time. Maintenance is an important consideration for analog magnetic recorders.

The earliest magnetic recorders were limited to a single channel of audio. Eventually, the ability to fabricate more than one gap in a head assembly opened up the ability to record more than one audio signal at a time. The alignment of the gaps is a critical aspect of the ability to interchange tapes between machines, so the alignment must be very accurate. The ability to create multiple gaps close together makes it possible to record many signals at the same time on the same piece of tape. The real functional breakthrough came, however, when it was realized that the ability to play previously recorded signals and record on other channels at the same time would make the process we know as **overdubbing** possible. Because the playback head is separate from the record head, signals reproduced from the playback head would not be in synchrony with the signal recorded at the record head. This issue would result in a time delay between old and new signals on tape. The solution is to allow playback on previously recorded channels through the record head, thereby keeping the playback and record gaps aligned.

Multitrack recording and synchronous playback/recording go together to allow the ability to add new material to tracks already recorded. Because the record and monitoring playback gaps are the same, the optimization of gap length for recording or playback must be sacrificed to allow both functions through the same gaps. Generally, the playback response through the record head is inferior to that of the dedicated playback head, but the best analog recorders have very good response even through the record head gaps. One problem inherent in overdubbing is the leakage of the record signal into the playback gaps, especially on adjacent tracks. The leakage is caused by the bias radiating from the recording gap carrying the audio signal into nearby playback gaps. This leakage does not affect the final playback – only the monitoring while overdubbing is under way.

The number of tracks that can be recorded depends on the width of the tape and the gap widths deemed sufficient for the desired signal-to-noise performance, as the recovered signal amplitude depends on the surface area of tape available to generate the magnetic field sensed by the head gaps. Standard reel-to-reel tape widths range from 1/4", 1/2", and 1" to 2" for open-reel recorders. Quarter-inch tapes are usually used for stereo (two-track) masters, although more tracks can be recorded on quarter-inch tape with reduced performance. Standard professional recorders use wider gaps and can accomplish 24 tracks on 2" tape, although 2" 16-track recorders produce somewhat better sound quality because the wider tracks produce more magnetic flux in the head gap. Recently, 1" two-track mastering machines have been created. At the other end, 8-track quarter-inch machines have also been produced. Because magnetic fields radiate into space, adjacent tracks can "bleed" if they are placed close together – another consideration in the design of multitrack heads.

The playback head gap acts as a scanning window as the tape moves over the head. It is able to sense the field near the head gap for short wavelengths, but as the wavelength grows, the entire head core begins to couple to the magnetic field. This behavior generates a series of peaks and dips in the low-frequency response, as distant parts of the tape field cancel and reinforce with the signal at the gap. This interaction is affected by the speed of tape movement, with bumps appearing at lower frequencies with lower tape speeds. Though high frequency and noise performance increase at 76 cm/s (30 ips), the head bump frequencies are also moved up an octave. Variations in head design contribute to significant differences in sonic performance between manufacturers' machines running at the same tape speed.

Because analog recorders' performance depends on a constant velocity of medium across the heads, the mechanical design of these machines is important. Tape is unwound from the supply reel, pulled across the head assembly, and wound onto the take-up reel. As the reels rotate, the tape is fed from one reel to the other, resulting in an increase in the rotation rate of the supply reel and a reduction in the rotation rate of the take-up reel. To keep the rate of tape across the heads constant, a capstan rotates at a constant speed and a pinch roller presses the tape against the capstan. The constancy of the capstan rotation provides a constant tape speed. Advanced designs have used servo control to allow operation without a capstan, but many analog recorders still employ the capstan to provide constant tape speed.

RECORDER ALIGNMENT

Due to the mechanical and electronic adjustments necessary to guarantee the best performance, analog recorders often require maintenance, including the frequent adjustment of critical electrical settings like bias and equalization for each channel and each tape formulation used. Further, because tapes are likely to be exchanged between machines, the physical alignment of the heads needs to be standardized. This standardization is accomplished by the use of standard calibration tapes containing full-track recordings of sine tones at different frequencies to allow electronic and mechanical adjustments. Before any use of calibration tapes, the tape transport and heads should be cleaned and demagnetized to prevent any damage to the tape. The calibration tape is used to adjust playback, and then the record heads and electronics are adjusted to produce a recording that mimics the calibration tape. Generally, calibration tapes are full track, not divided into specific channels. This design allows one tape to be used to calibrate all recorders of a specific tape width regardless of how many audio tracks are actually recorded. The drawback to this approach is that because there is magnetic flux on the tape in what would ordinarily be guard bands between tracks, the low-frequency playback signals are larger than they would be if the tape had been recorded on the machine being calibrated. Calibrations must then be adjusted by an amount specified by the manufacturer for low-frequency playback. This increase in low-frequency signal is due to the fringing effect (the magnetic field near the gap being sensed by the head).

Adjusting the bias level gives the engineer some creative options in selecting the balance of noise level, distortion, and high-frequency sensitivity. Each tape formulation has a different set of curves relating these factors to the bias signal level. The amount of harmonic distortion and the sensitivity of the tape are strongly influenced by the amount of bias signal. The maximum output levels and modulation noise also depend on the bias level. If we feed a 10 kHz sine wave into the recorder and increase the bias level from zero, we will see a peak in the output level indication as we raise the bias level. If we further increase the bias, the output level will decrease. Often the manufacturer will recommend a bias level that produces 3 dB of reduction from the peak, an overbias of 3 dB. The bias adjustment must be made on each channel separately, a rather time-consuming endeavor on a 24-track machine.

To further complicate the calibration process, several different equalization curves have been used in the design of analog recorders. A separate calibration tape is required for each curve and each tape speed. Most manufacturers recommend a particular equalization system, such as NAB or IEC (CCIR). Often a tape formulation is adopted by a studio because it delivers the sound desired and minimizes the amount of recalibration needed between projects. Some later recorder designs provide automatic calibration, but tedious manual setting of the many equalization (EQ) and bias adjustments is more common.

The three-dimensional adjustment of the heads relative to the tape is important (Figure 8-8). The height and tilt of the heads relative to the tape is adjusted to guarantee the longest head life and best electronic performance possible. The height adjustment determines how the head gaps align across the width of the tape as it is pulled over the heads. The zenith adjustment controls how the heads lean outward: neither the top nor bottom should lean farther out, so that the heads will wear evenly over time. The alignment most often requiring adjustment is the azimuth – the angle the head gaps make with the longitudinal axis of the tape. If the azimuth is misadjusted, the top and bottom tracks will not be reproduced in proper time alignment when played on a properly adjusted machine. Frequency components exactly half a wavelength apart will cancel when summed. For 20 kHz sine

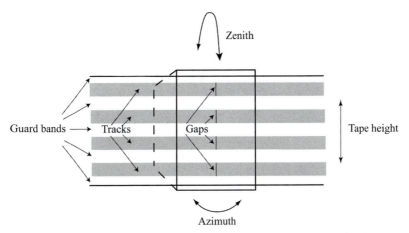

Figure 8-8 Head alignment parameters. Azimuth adjusts gaps perpendicular to tape travel.

waves, this amounts to less than a 10 μm difference! Azimuth is adjusted by playing a full-track calibration tape while changing the setting until the outermost tracks are in phase at high frequencies. This task may be observed on an oscilloscope as an x-y plot of the two outer channels in which proper alignment shows as a line of slope $+1$. This procedure should be started with lower-frequency input signals, increasing the frequency so that the azimuth is not adjusted off by a full cycle at higher frequency.

There is no question that digital recording has become the dominant technology for sound recording. Still, many of the professional tape recorders in use in the heyday of analog recording are finding their way into the hands of eager engineers who realize the contribution that analog technology can still make to music recording. Analog tape and recorders are now in the hands of a few dedicated entrepreneurs continuing to improve the technology.

SUGGESTED READING

Ballou, G. M. (2002). *Handbook for Sound Engineers* (3rd ed.). Focal Press. ISBN 0-240-80454-6. See Chapter 28, Magnetic Recording and Playback.

Bertram, H. N. (1992). *Theory of Magnetic Recording*. Cambridge University Press. ISBN 0-521-44512-4.

Mee, C. D., & Daniel, E. D. (1990). *Magnetic Recording Handbook*. McGraw-Hill, Inc. ISBN 0-07-041274-X.

Woram, J. N. (1989). *Sound Recording Handbook*. Howard W. Sams & Co. ISBN 0-672-22583-2. See Chapters 9 through 11.

CHAPTER NINE

Digital Audio Recording and Processing

Contents

Computers have changed the process of sound recording rather dramatically. Where we once relied on an interconnected system of separate electronic devices, we can now accomplish the same tasks using computer software – once we have converted our audio signals into computer data. There are fundamental differences in how computers and analog audio systems behave that we must understand in order to take full advantage of the capabilities of digital audio systems. We can start by considering how computers represent numbers.

THE BINARY NUMBER SYSTEM

In order to make sense of digital audio devices, we need first to understand how binary arithmetic is used to represent the signal and manipulate it mathematically. Like the decimal system, **binary digits (bits)** increase in weighting as we move leftward in the word from least significant bit (LSB) to most significant bit (MSB). See Figure 9-1.

We are used to the decimal system, a number system based on tens. Each decimal digit represents a multiple of ten. In binary arithmetic, each digit represents a multiple of two. Computers use binary mathematics, but we are

The Science of Sound Recording
ISBN 978-0-240-82154-2

137

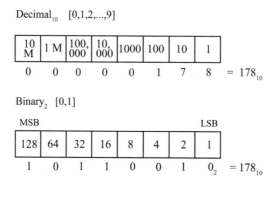

Figure 9-1 A comparison of number systems. The base ten or decimal system is most familiar. Computers use binary and related octal and hexadecimal systems.

able to program them using decimal numbers that are converted to binary by the software. When acquiring data from an analog audio signal, however, the conversion produces binary numbers directly. Number systems using 8 and 16 as the base sometimes make handling digital information simpler; these systems, called octal and hexadecimal, are frequently encountered in software. Octal representations are denoted by the subscript "8" and hexadecimal by "H":

$$11111112 = 1778 = 7FH = 12710$$

Although they execute code in binary, computers can be programmed using octal or hexadecimal numbers and the registers read back from digital hardware often display the results in these forms. Both octal and hexadecimal are easily translated into binary; decimal is not as easily converted. (You'll feel much younger if you say you're 28H as opposed to 4010!)

A number system should accommodate negative numbers as well as positive. The solution often used with binary is called **two's complement** (Figure 9-2). In two's complement, the MSB indicates the sign (although it retains its weighting), positive or negative. The two's complement is computed by subtracting the number from $2n$ where n is the number of bits in the word. Positive numbers are used directly using zeros to fill out the leftmost bits if necessary. For negative numbers, each bit in the word is

Binary	Unsigned	Signed
000	0	0
001	1	1
010	2	2
011	3	3
100	4	-4
101	5	-3
110	6	-2
111	7	-1

Figure 9-2 Three bit binary words for signed and unsigned integers. Two's complement is used to represent signed numbers.

inverted (complemented, equivalent to subtracting from $2n - 1$) and 1 is added to the LSB. Overflow from the MSB is ignored. Once in two's complement form, the addition operation works without consideration of sign.

COMPUTERS AND TIME

Our experience of the world is one of continuous phenomena. Continuity implies that no matter when we make a measurement, there is a distinct value associated with that point in time. As we have seen, this is sometimes an illusion due to our scale of observation, as is the case with sound propagation through a gas. The sound pressure that appears to us to be continuous is in fact due to the sum of many individual molecular collisions, each contributing a tiny amount of force. By summing so many of these events, we can use continuous functions to describe the observed behavior of physical systems.

Computers do not function as continuous systems. Each program instruction takes a finite amount of time to execute. Each measurement takes time to make. Fortunately, we can take advantage of our perception of time, allowing computer processes operate as seemingly continuous systems if their speed is sufficiently faster than our sensory systems. Further, it can be proven mathematically that even if our measurements are not continuous,

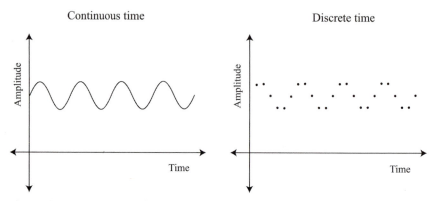

Figure 9-3 Continuous and discrete representations of the same sinusoidal signal. Though the discrete representation is sampled, a sine wave is still discernable.

we are able to determine the continuous values as long as our measurements are made quickly enough relative to the rate of change of the quantity under observation. A few key equations form the foundation of sampling theory, allowing the reversible conversion from continuous to discrete representations of our measured quantities. Figure 9-3 shows these two representations for comparison.

DIGITAL AUDIO: THE THEORY

The sampling process converts continuous signals like audio signals into a series of discrete numbers that measure the voltage of the signal at a fixed time interval T corresponding to a sample rate of $f_s = 1/T$. As is so often the case, there are problems implementing the ideal theoretical approach using actual electronic devices. Examining the theory will help us to understand why in practice such conversions may not be executed without some alteration of the signals.

The **Shannon sampling theorem** states that as long as a function $f(t)$ contains no frequencies (at or) above W Hz, it is completely determined by its ordinates at a series of points spaced $1/2W$ seconds apart. More simply, the sample rate must exceed twice the highest sinusoidal frequency component of the signal. It is worth noting that the theorem, proposed by Claude Shannon, Harry Nyquist, and others working in information theory, was devised long before computers existed that could take advantage of the theorem. The sampling theorem is often called the **Nyquist theorem**, and the $f_s/2$ frequency is known as the Nyquist frequency.

As sampling hardware became available, the difficulty of implementing the rather simple theory quickly became evident. The theorem assumes that the sampling period timing and measurement of the signal voltage are perfectly executed, and that is where the complications begin.

Analog audio signals are voltages as a function of time. When using analog electronics that produce a continuous signal, this design is sensible. Once we sample the data, we are able to convert it to voltage as a function of frequency. This approach is similar to how our ears transmit sound information to the central nervous system. Though it could be done with analog filters in theory, it would be highly impractical. The voltage versus time representation is known as the **time domain**; voltage versus frequency is known as the **frequency domain**. Figure 9-4 compares these domains in graphical form.

Conversion between the time domain and the frequency domain is accomplished using the **Fourier transform**. The Fourier transform and its opposite, the inverse Fourier transform, allow data to be represented in either form, simplifying many operations needed for digital signal processing. The power of the Fourier transform concept is revealed in applications such as spectrograms that plot frequency against amplitude in real time, changing as the signal changes. Separating, grouping, and manipulating the spectral components individually allow the adjustment of the tuning of one instrument in a mix or tuning one string of a recorded guitar. Background noise can be identified and eliminated. Hidden sounds masked by much louder ones can be revealed. These processes depend on the ability

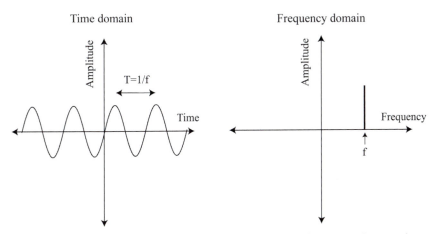

Figure 9-4 The same sinusoidal signal in time domain and frequency domain plots.

to isolate the spectral components of a signal offered by the Fourier transform, which may then be edited and recombined.

The Fourier transform equation is a sum of sinusoids. The transform is based on the **Fourier series** (Equation 9-1), an infinite sum of cosines that can be used to describe any periodic signal:

$$\varphi(\gamma) = a_0 \cos \frac{\pi \gamma}{2} + a_1 \cos 3 \frac{\pi \gamma}{2} + a_2 \cos 5 \frac{\pi \gamma}{2} + \dots \qquad \text{9-1}$$

The a_n coefficient gives the amplitude of each frequency component; the cosine term gives the frequency of that component. Because it is an infinite series, it will not provide an exact value unless we have an infinite number of terms, which we never do in practice. In order to make use of the Fourier series, we must find a method of using a finite number of terms to derive the answer. The **discrete Fourier transform** (DFT), Equation 9-2, is such a method of using the Fourier series in a practical way:

$$X(\omega_k) = \sum_{n=0}^{N-1} x(t_n) e^{-j\omega_k t_n}, \quad k = 0, 1, 2, \dots, N-1 \qquad \text{9-2}$$

The function $X(\omega_k)$ is the complex frequency domain transform of time domain function $x(t_n)$. ω_k is the kth frequency sample of the frequency domain transform (where k steps between 0 and the number of samples minus one); $x(t_n)$ is the signal amplitude measured at time t_n (where there are n steps between 0 and the number of samples minus 1) and N is the number of samples. The equation sums the sampled value $x(t_n)$ scaled by the complex term $e^{-j\omega_k t_n}$ for each sample. It is expressed in Equation 9-3 using a convenient mathematical identity to simplify the expression, namely, Euler's identity:

$$e^{j\theta} = \cos(\theta) + j \sin(\theta) \qquad \text{9-3}$$

The complex exponential thus gives the complex sinusoidal frequency at each sample, providing both frequency and phase information for each sinusoidal component. Equation 9-2 provides an algorithm that can be optimized for computer execution. What we should realize from these equations is that any periodic signal can be reduced to a sum of properly weighted sinusoidal functions.

The DFT and other transforms that do not make use of an infinite number of coefficients are susceptible to the **Gibbs phenomenon**. This effect causes the transformation to overshoot the actual value it should compute, allowing outputs that may exceed the range of the hardware. The ringing is due to the finite number of terms failing to reach the limit of the infinite series on which the approximation is based. The overshoot occurs at what are called **jump discontinuities**, an example of which occurs at the rising and falling edge of a square wave signal. There, the failure to converge to the limit reached at infinity causes the sum of terms to exceed the correct value. The effect is potentially important when we digitally limit a signal to 0 dB FS using a software limiter. Setting the limiter maximum output to 0.5 to 1 dB below full scale will prevent the overshoot from exceeding the maximum value of 0 dB FS. Because the overshoot may appear in the output between samples, it must be detected by software that looks at the intersample output level.

SAMPLING: QUANTIZATION

At fixed intervals (the sampling period T), the signal voltage is measured in a process known as **quantization**. Quantization converts a continuously varying voltage into a series of discrete measured, or quantized, binary numbers. The digital representation of the analog signal is encoded using digital words, with the number of bits per word determining the number of discrete voltage values we can discern in the resulting data. The more bits per word, the finer the minimum difference we can distinguish. The smallest level change we can encode is the analog level that corresponds to the difference between the least significant bit being set on or off. If we fix the maximum input signal amplitude, the minimum difference we can resolve will determine the overall dynamic range we can provide.

In analog audio, we are accustomed to thinking of the low amplitude limits of the system being determined by the residual noise in the system. The noise level is determined mainly by thermal noise in circuit elements and the noise sources of the active devices employed. Thermal noise has a constant power density across the frequency spectrum, producing the familiar white noise. This noise is not an absolute limit, however, because we can still hear a signal smaller than the noise until it is fully masked by the noise. In digital audio, the low-amplitude limitations are due to the error in measuring the small signal resulting from the finite number of bits in the word length we use to encode the signal. As the input signal level decreases, the error in the measurement relative to the signal amplitude increases. With large signals, many bits are available to describe the signal amplitude. As the signal level decreases, it approaches the voltage corresponding to the LSB in the converted word. The difference between the signal voltage and the LSB voltage equivalent is the error in the conversion. When the signal falls below the LSB equivalent voltage, the output may even be zero. The error signal is correlated with the signal, unlike the constant wideband noise of analog electronic systems. The signal may cut in and out as it drops below the LSB threshold and returns. The signal-to-error ratio decreases as the signal level decreases.

To overcome the signal-correlated error problem, a small amount of wideband noise called **dither** is added to the signal. The random nature of the dither decorrelates the quantization error from the signal, producing a more natural-sounding noise floor. Although adding noise would usually be undesirable, there is an optimal amount that decorrelates the signal from the error at levels well below audibility. Optimal dither is in the range of 1/3 to 1 LSB equivalent depending on the type of dither used. Different noise spectra produce different-sounding dithered signals although the difference is small when listening to full-level signals. Dither must be added whenever word length is shortened in digital audio computations as well as during the initial sampling.

The dynamic range of a digital system is approximately equal to 2 to the power of the number of bits, n, in the word, or $20 \log 2^n$. More accurately, the dynamic range of a digital system using n bits per word is:

$$SNR(dB) = 6.02n + 1.76 \qquad\qquad 9\text{-}4$$

Because the energy in the error produced by the converter is a statistical probability function, the error must be calculated by integrating the product of the error and its probability. We can use the $20 \log 2^n$ approximation to see roughly how word length affects the dynamic range of digital audio systems in Figure 9-5.

In theory, the more bits used to encode sample words, the greater our confidence in the accuracy of the measurement and the better the sound quality. In reality, the physical process of conversion limits the real accuracy predicted by theory because real converters fall short of theoretical perfection. Analog to digital converters claiming 24-bit quantization in reality convert only 20 to 22 bits linearly. Nonetheless, the more bits we use, the better the fidelity – as long as the electronic circuitry used can perform with the required accuracy.

n Bits	Dynamic Range (dB)
2	12
4	24
6	36
8	48
10	60
12	72
14	84
16	96
18	108
20	120
24	144

Figure 9-5 The number of bits in a word determines the dynamic range of a digital system. Dynamic range increases by roughly 6 dB per bit.

SAMPLING: TIMING

The other critical factor in our measurement of the analog signal is the regularity of the sampling. We assume that each sample is taken at precisely the same interval. Small deviations in the timing will result in sample error, because the signal is changing with time and measurements made at incorrect times will have different values than samples made at the correct time. Sampling requires very stable timing reference and that the circuitry is able to perform the conversion within the time allowed between samples. Variations in the sample time are known as **jitter**, a potential cause of error in the conversion process.

The sampling theorem predicts that a continuous function, like an analog audio signal, can be exactly represented by a sampled discrete time sequence, as long as the sample rate is greater than twice the highest frequency contained in the original signal. The theory assumes an infinite discrete time sequence, however, which is impossible to accomplish with real converters. Nonetheless, if the signal spectrum is limited to less than half the sample rate, we are able to reconstruct the original signal adequately.

In motion pictures, the result of sampling at too low a rate is obvious: the spokes of a rotating wheel appear to spin backward when the wheel rotates more than one rotation between frames of $1/24$th second. Instead of rotating backward, the frequencies above half the sample rate are "folded over," creating audible frequencies that fall below half the sampling frequency. The dots in Figure 9-6 mark the sample times, showing how the high-frequency sine wave has undergone more than one cycle between samples. The resulting sine wave is of a lower frequency than either the signal or the sample rate. Alias frequencies are produced as sums and differences of the sample rate, and the audio frequencies present above half the sample rate. Only the difference alias falls within the audible frequency range.

In order to prevent aliasing, analog low-pass filters are required at the A/D (analog/digital converter) input to remove any frequencies greater than half the sample rate. For 44.1 kHz sample rates, this means that frequencies above 22.05 kHz must be greatly attenuated. Analog filters capable of very sharp corner frequencies are quite difficult to implement, making such converters complicated. Each stage of the filter produces 6 dB/octave attenuation, so many stages are required to reduce the high-frequency content above half the sample rate dramatically without reducing frequencies below the critical frequency. These filters may degrade the audio signal audibly, as such complex analog filters introduce phase and amplitude irregularities at frequencies well below the cutoff.

Once the sampled data is acquired, it can be processed using digital filtering. The analog antialias filters may be replaced by digital filtering in a process known as **oversampling**. The sample rate is set at several times the ultimate desired rate, and the excess data is filtered digitally to produce a lower sample rate through a process known as **decimation**. Decimation averages the samples to produce samples at a lower effective sample rate. Decimation allows relaxed low-pass filtering of the input because it raises

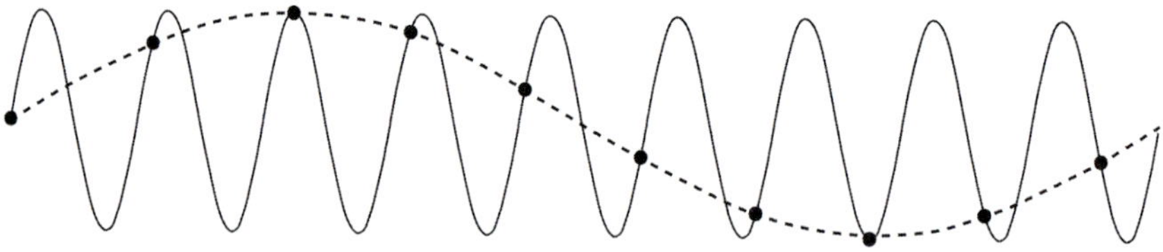

Figure 9-6 A simple example of aliasing. The alias frequency created is the difference between the signal frequency and the sample rate.

the input sample rate F_s so that less-steep analog filters are still able to limit the signal frequencies to below half the sample rate. Further filtering is accomplished digitally when the data is decimated. Decimation allows an elevated sample rate without generating more data.

In theory, the sample rate must be only higher than twice the highest frequency present in the signal. How much higher it must be in practice is open to discussion. The 44.1 kHz sample rate used in compact discs and many other digital devices was established at a time when digital storage space was limited, in both delivery media and recording devices. We are now able to handle much higher data rates and can use higher sample rates and longer words, but it is not clear how much is enough. The balance between high sample rates and the amount of data that needs to be stored is a trade-off that must be decided by digital recording system users.

ANALOG-TO-DIGITAL CONVERSION

Because it takes a finite time to measure the continuous signal, the signal can change between the start and end of a conversion cycle. To prevent this change, a circuit called a **sample-and-hold** (S/H) is used to hold the sample voltage until it can be measured. This hold can be done with a capacitor and an electronic switch to allow the capacitor to charge to the signal voltage and then disconnect from the signal, holding the final voltage. **Analog-to-digital converters** can be designed in several ways. Originally, digital devices used multibit converters, converting each word by comparing the signal to a digital-to-analog converter's output and continually adjusting the digital word that sets the D/A converter's output until the two voltages agree. This process is known as **successive approximation** conversion (or "guess-and-check"). Resolving each bit in the data word requires an iteration of the algorithm. Modern A/D converters use **sigma-delta conversion**, in which the signal amplitude is compared with its previous value at a very high rate and a single bit is used to tell whether the latest measurement is larger than the previous one. These individual differences are then summed to produce the multibit digital word. The one bit data can also be used directly, without conversion to multibit words. This system is known as **delta modulation**. The delta modulation data stream is very simple to convert to a continuous signal, as only integration and a gentle low-pass filter are required. Direct-Stream Digital (DSD) is a form of delta modulation used for SACD (Super Audio Compact Disc) encoding. Professional DSD has begun to been accepted as a viable alternative to the analog two-track as a mastering format. Each of these systems has strengths and weaknesses. Multibit digital data can be used directly for digital signal processing that is not possible with the one-bit data stream. Multibit conversion is more widely used due to its processing flexibility.

A/D converters ideally produce output words that are linearly related to the input voltages they measure. In some cases, however, there are deviations from linearity, especially in the low-order bits. Some codes may be missing or the voltage steps may not be exactly equal. These faults limit the accuracy of the conversion and may produce audible distortion of low-level signals. Changes in the temperature of the converter circuitry can contribute to drift over time. With careful design, these errors can be minimized. Most modern converters produce acceptable linearity over a wide amplitude range, although 24-bit converters are not likely to produce real 144 dB dynamic ranges, in part because the analog circuitry cannot produce such low noise levels at room temperature. Chapter 7 discusses thermal noise limits in analog audio systems.

DIGITAL-TO-ANALOG CONVERSION

Sampling in the time domain results in modulation of the signal by the sampling pulse, effectively multiplying the two signals. Once sampled, the data then contains not only the original data but also multiples of it reflected about integer multiples of the sample rate (Figure 9-7). These multiples are

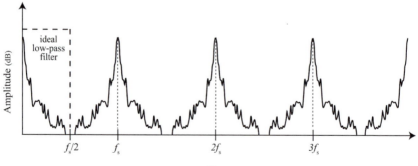

Figure 9-7 Sampling creates multiple images of the original spectrum wrapped around multiples of the sample rate. Removing the images requires a low-pass antialiasing filter.

called **images** of the original signal spectrum. Before the data can be converted back to a continuous signal, all but the original spectrum must be removed. Like the analog-to-digital conversion, the digital-to-analog conversion involves low-pass filtering of the data, although for a different reason. The same issue of constructing a sharp filter exists for the D/A conversion as it did for the A/D conversion. Here again, most of the filtering may be done on the digital side.

When viewed from the frequency domain, each sample represents the sampled signal convolved with the sample pulse. **Convolution** here is the frequency domain equivalent of the multiplication of the signal by the sampling pulse that occurred in the time domain. The sampling pulse approximates an **impulse**, an infinitely narrow pulse. (Mathematically, an impulse, or δ function, has an amplitude of 0 everywhere except at the origin, x = 0, where its value is 1.) The response of any system to an impulse excitation is its **impulse response**. Digital samples are mathematically treated as scaling coefficients of impulses at each sample time. The **sinc function** (Equation 9-5) is the frequency domain representation of the sampling pulse. Signal processing theory provides that the sampled signal can be restored to a continuous form through a summing of sample-weighted sinc functions. When they are integrated, their output reconstructs the original sampled signal plus images of the spectrum wrapped around multiples of the sample frequency.

$$\mathrm{sinc(t)} = \frac{\sin(\pi t)}{\pi t} \qquad\qquad 9\text{-}5$$

In order to properly reconstruct the original data, the stored measurements must be read out as infinitesimally narrow pulses of varying heights that could then simply be low-pass-filtered to restore the continuous signal. Such narrow pulses are not possible to create with electronic circuits, so instead the data values are held until the next sample time using a sample-and-hold circuit. The held values form the familiar stairstep appearance of the unfiltered output. The output is then low-pass-filtered. Because the time domain low-pass filter is equivalent to the **sinc function** in the frequency domain, the conversion from discrete to continuous representation takes advantage of the sinc function. The sinc function, presented in Equation 9-5, is the normalized function used in signal processing. It is normalized by π so that zero crossings occur at integer values along the x-axis. The unnormalized sinc function $(\sin(x))/x$ is plotted in Figure 9-8, with zero crossings at multiples of $+/- \pi$.

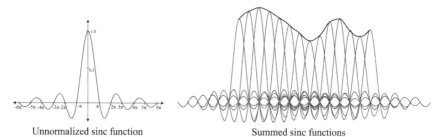

Unnormalized sinc function   Summed sinc functions

Figure 9-8 The sinc function plotted. A sinc function represents the sample data from each sample. Summing sequential sinc functions reproduces the continuous waveform.

The sampling process results in data that may be recombined into a continuous analog voltage by summing sinc functions for each sample value. Low-pass filtering completes the conversion to continuous by smoothing the output voltage and eliminating multiples of the original signal spectrum created by the sample pulse multiplication in the A/D step.

Notice that the sinc function extends into the negative x-axis. It is possible for digital filters to preing when excited by a signal. No, they do not go backward in time: the output is delayed so that the preing actually occurs when the filter is first driven, before its response peaks. Time delay is both an advantage and a disadvantage of digital audio processes. In the analog circuit world, creating delay is complicated. It is easy with digital systems, but it is also unavoidable, and it introduces a concern not encountered in the analog studio: time synchronization. All devices participating in a digital audio system must be clocked together. All software used in digital mixing must execute in synchrony or be resynchronized to avoid phase cancellations when originally simultaneous signals are combined after processing.

Just as multiplication in the time domain becomes convolution in the frequency domain, the reverse is also true. Thus the ideal filter to remove images in the frequency domain, as shown in Figure 9-7, a rectangular low-pass filter that encloses only the frequencies of the original signal, has a sinc function in the time domain. The result of this filter in the time domain looks like a sinc function, hence the name "sinc filter." The symmetry of the sinc function about the y-axis and its infinite extension complicate its use in digital filter programs, so in practice it is windowed to reduce its span, using only the positive coefficients. Smoothly limiting the extension of the sinc function to avoid the filter ringing caused by simply truncating the sinc function, the data are multiplied by a bell-shaped windowing function. Proper choice of the window function shape creates a sharp attenuation just above the cutoff frequency with little ringing below the cutoff.

Another issue with D/A converters relates to what happens as the digital word feeding the converter changes from sample to sample. The output of the digital-to-analog converter (DAC) is always connected to the analog output, so any glitches that occur as the digital word changes would be sent to the output. To prevent this, a sample-and-hold circuit is used. The S/H circuit samples the analog value at the DAC after any glitches subside and stores it on a capacitor until the next value is ready. This is an analog sampler, but it is still a form of sampling. The width of the sample-and-hold pulse affects the frequency response of the system and can be used to tune the output frequency response of the converter to offset other aspects of the system that may alter the response, this effect is known as the **aperture effect**.

Both A/D and D/A converters can be improved by running at much higher sample rates than the ultimate sample rate and using digital signal processing to restore the desired output sample rate. This technique, known as **oversampling**, has one big advantage: the ability to use digital filters instead of steep analog filters to remove aliasing and imaging. In the case of A/D conversions, the quantizer can be run at much higher sample rates than the final sample rate and the data averaged to produce the lower sample rate, a process known as **decimation**. At the D/A end, analog values at times between stored samples can be created using a process known as **interpolation**. Interpolation filters work by inserting multiple, evenly spaced zero-valued samples between sampled values in the data stream, then adjusting the sample values by a series of coefficients derived from the sinc function. Interpolation allows sample rate conversion to higher sample rates, or **upsampling**, on data already sampled.

Multibit converters are most widely used in digital audio devices. These digital systems employ coding schemes that make use of a multibit digital word to encode the measured analog signal amplitude. The coding system used is most often **pulse-code modulation** (PCM), in which the bits' states, on or off, code for 1 and 0 in the binary digital word. Other methods can use the width of a digital pulse to encode amplitude in a system, known as **pulse-width modulation** (PWM), or the distance between pulses, known as **pulse-position modulation** (PPM). The advantage of the PCM system is that it can easily be used for calculations by computer logic circuitry. The newer DSD data cannot as easily be manipulated, although it can be stored.

DIGITAL AUDIO INTERCONNECTION

Sending digital audio between devices can be more involved than connecting analog devices. In addition to providing the data, the connection must transmit clock information so that the received data may be played or processed at the proper sample rate. The data words can be transmitted all bits at once or one after the other, in parallel or in series. The requirement for many separate wires necessitated by a parallel transmission makes cables expensive; hence serial transmission is preferable as long as the bits are transferred quickly enough. Parallel connections have largely been replaced by fast serial transmission systems.

For sampled audio data, there are two common two-channel serial protocols: S/PDIF (Sony/Philips Digital Interface) and AES3, also called AES/EBU. The data protocols are similar; however, S/PDIF uses low-voltage, unbalanced 75 Ω coaxial wiring, and AES/EBU uses a higher-voltage, balanced 110 Ω XLR connection. S/PDIF is also available as an optical interface. In a serial data transmission, the beginning of a word must be indicated by a special code in the data stream. Both formats transmit sample rate information as well as other information in addition to the audio data in what is known as **subcode**. The data is protected from transmission errors by redundancy encoded in the transmitted signals. Both protocols are self-clocking, so a separate clock line is unnecessary.

Multichannel data may also be transmitted. MADI (multi–channel audio digital interface), or AES 10, uses coding similar to AES3 to distribute up to 56 channels on coaxial cable. MADI runs at a fixed data rate regardless of the sample rate, and although it has self-clocking information embedded in the data streams, it also uses a dedicated master synchronization wire. Digital audio can also be sent over Ethernet, a technique that is gaining popularity. Many proprietary multichannel protocols have been devised, such as Tascam's T-DIF and the Alesis ADAT lightpipe. These older protocols are still in use, but have been relegated to consumer and "prosumer" system uses due to their low cost but limited capabilities. The eight-channel ADAT interface uses separate fiber-optic connections to transmit and receive. It sends 24–bit words but is limited to 48 kHz clock rates. (Higher sample rates are possible by using two tracks for each channel.) The T-DIF interface is 8 channels at 24 bits and similarly limited to a 48 kHz clock rate. It is bidirectional, using a single cable multiconductor cable. Although both formats are proprietary, the ADAT interface has gained popularity due to its low cost.

DIGITAL RECORDING

Whatever system is used to acquire the digital data, we are faced with the same dilemma we encounter in the analog recorder when it comes to storing the information. Although computer memories may store the data temporarily, most computers use **dynamic random access memory** (DRAM) chips that lose the data when the power is removed. For long-term storage, magnetic media are used for digital recording as well as for analog. Because digital data require only two states and not the complete linearity demanded by analog recordings, there is a difference in the way the process is applied. Digital magnetic recorders use saturation recording, leaving all magnetic domains polarized in one direction or the other with no intermediate levels required, so the bias current used in analog recording is unnecessary in digital recorders. The density is quite high in digital recorders, which introduces some problems not encountered in analog recorders. The number of bits per unit area of medium is limited in the longitudinal recording method used for analog record, although it is sufficient for the demands of analog recording. Digital magnetic media may benefit from closer packing of domains, which is achieved by using perpendicular recording, in which the domain magnetic fields are magnetized perpendicular to the medium surface instead of along the surface as in analog recorders (Figure 9-9).

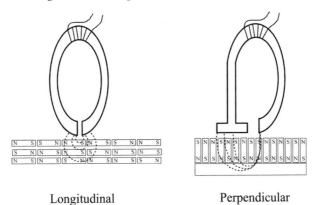

Longitudinal   Perpendicular

Figure 9-9 Comparison of longitudinal recording used in analog magnetic recording and perpendicular recording used in many disc drives.

Digital data stored magnetically requires only two discernable states for binary information. This requirement is easily achieved by magnetizing domains fully in one or the other polarity. Though this approach avoids the nonlinear region of the M-H curve, it introduces another problem: the interference between closely occurring bits. If the write head and the medium are not capable of altering the magnetic polarities as rapidly as the bits are changing, the magnetization from the previous bit will affect the next bit, which causes the data to be altered because the overlap makes discriminating between ones and zeros unclear. This intersymbol interference limits the data density that can be stored.

We have several options for storing digital data, including dedicated devices using tape or discs as media and general-purpose personal computers with added interfaces to acquire and convert analog audio. The high data density required for storing digital data made early digital recorders quite complicated, requiring rotating magnetic head recorders designed for video recording or using high tape speeds with stationary heads requiring more than one data track for each audio channel to provide enough bandwidth. The personal computer has largely replaced the mechanically complicated digital recorders as the preferred storage device for digital audio recordings. The low price and wide availability of large, fast disc drives has spurred a move to the computer as the digital audio recorder of choice, especially as the computer can take on the functions of editing, mixing, processing, and storing the entire project in a single device. A recent development is the flash RAM chip, popular in USB memory sticks for example. This magnetic nonvolatile RAM – though slower to read and write than a hard disk – is becoming popular for non-time-critical recording such as backup of sound files and stereo sound file distribution.

The evolution of digital recorders has been rapid. Rotary-head modular digital multitrack recorders and stereo DAT recorders enjoyed only a few years of widespread use before the move to the general-purpose computer as the preferred platform for digital recording. These machines temporarily bridged the gap between high-cost stationary head professional digital recorders like the Sony DASH and Mitsubishi Pro-Digi systems and analog multitrack. The Alesis ADAT and TASCAM DTRS machines used videotape, which was cheap and readily available, to provide inexpensive access to digital recording for a wide range of users. These machines, though initially inexpensive, suffered from their complexity when head wear and transport malfunctions required difficult repair and diagnosis procedures. Yamaha produced the DMR/DRU series of recorders, which used stationary heads and proprietary tape cassettes, to deliver 20-bit 8-channel digital recording in the early 1990s, but they were expensive relative to the ADAT and DTRS machines and never caught on. None of these tape-based systems survived the move to computer-based systems, and all have been phased out or will soon be retired. Although tape provides the advantage of removable media, the large size of hard drives and the availability of plug-and-play computer interfaces for storage media has diminished the attractiveness of tape-based digital recorders.

The ability to use inexpensive, mass-produced personal computers for digital audio recording and mixing has greatly expanded the accessibility of these tools. The addition of a FireWire, USB, or Thunderbolt audio interface and some software is all that is required to create a digital studio entirely within the computer. This change has had a dramatic effect on the recording studio and the music business in general. Essentially, the entire recording studio can now be contained in a single piece of equipment, with the ability to recall the entire project and studio configuration in a few seconds. The advantages of digital audio are hard to ignore, even for those dedicated to the analog studio paradigm.

Use of personal computers for audio recording has introduced a new set of difficulties. Each operating system and hardware platform requires different software, and there are differences in the bus structures and interface ports available that complicate the choice of peripheral audio interfaces (Figure 9-10). Input/output buses include FireWire and USB high-speed serial interfaces, both of which are possible choices for connecting multi-channel A/D and D/A modules to the computer to provide audio access. The Thunderbolt interface protocol promises even faster device interconnection. The software for recording interacts with the operating system to access these audio inputs and may do so with differing speed capabilities on different computers. With the main choices for personal computer operating systems – Macintosh OS X, Windows, and Linux – several types of interface are supported, but different recording programs are required and the performance of the audio interfaces may differ due to differences in the hardware and device drivers employed in the particular computer used. Recording engineers must now have some knowledge of the internal

FireWire 400	USB2	FireWire 800	USB3	Thunderbolt
400 Mb/s	480 Mb/s	800 Mb/s	4.8 Gb/s	10 Gb/s

Figure 9-10 Computer interface speeds.

workings of their computer. If something goes wrong with the recording system, it becomes necessary to isolate the problem by troubleshooting a complicated series of interactions between software, computer, and peripheral hardware that may not be well documented. Each manufacturer provides information about their part of the system, but no one company is responsible for the entire system, leaving the user to deal with the problem.

An issue we encounter with digital audio that is not found in analog systems is related to the time it takes to execute instructions. Even in complicated analog systems, computations occur in real time or instantaneously to human observers. Digital processes take varying amounts of time to complete, making parallel processes no longer synchronous. When monitoring inputs through computer-based audio systems (**software monitoring**), there is a time lag between the sound input and the sound played back by the program. The delay is a function of the sample rate and buffer sizes chosen. An alternative to this software monitoring is to monitor the analog inputs through a mixer at the input rather than through the software. Many systems include **automatic delay compensation** to resynchronize internal processes, but this function does not eliminate the delay we encounter from the A/D and D/A conversion processes. When digital audio devices are connected, their clocking must be identical to maintain synchrony. Thus digital audio introduces a need for clock distribution that is not found in analog systems.

One of the advantages of software is the ability to refine and upgrade its performance over time. This ability can also be a drawback if the compatibility issues we encounter with continual updates continue to render code obsolete at a rapid rate. Not only the inherent performance of the software itself must be considered but also the interaction of the applications with the operating system used by the computer. The operating system (OS), the code that determines the operation of the CPU and peripherals, is developed either by the company that makes the computer or by an outside company that provides the OS software. The recording application software may be written by programmers without advanced knowledge of the new developments in the OS. Keeping the OS and application software synchronized can therefore become a major issue. A full-time studio technical staff often provided such maintenance in the past, but the personal computer–based studio is frequently the responsibility of a much smaller staff or simply the engineer alone. The engineer must therefore become a knowledgeable computer technician in order to keep computer-based recording systems working smoothly.

DIGITAL SIGNAL PROCESSING

Use of a discrete representation of audio signals allows them to be processed using frequency domain processes to perform the tasks previously accomplished with analog hardware – and many tasks that previously could not be accomplished. Filters created from capacitors and inductors can also be simulated mathematically. Because sinusoids are common functions, many well-studied methods exist to manipulate them.

Convolution is a method of combining two signals to generate a new signal that contains features of both original signals. For example, recording an impulse sound such as a balloon pop obtains a room's acoustical characteristics from its impulse response. The impulse response is then convolved with an audio signal to produce the resulting reverberation. Impulse responses can be used to create a digital filter with the same characteristics. Convolution can also be used to examine the periodicity of signals through autocorrelation, a process that finds similarities between a signal and delayed versions of itself. Different signals can similarly be compared through cross-correlation to find similarities.

Simply by combining the current sample value with previous samples, digital filters can be implemented. The single sample delay is called a **unit delay**. The sample values can be scaled and summed. Arranging multiple delays in series and parallel combinations allows complicated filter shaping. This technique provides many audio processing operations including filtering and reverberation simulation. Cascading multiple delays produces a filter known as a **transversal filter**, in which the output depends only on past and current data values. This filter is also known as a **finite impulse response** (FIR) filter because it has an impulse response that does not extend forever. The FIR filter is said to be **causal** because its output depends on existing data only, not on future data. By using feedback within the filter algorithm, a method known as **recursion**, an **infinite impulse** (IIR) filter can be produced. These filters have an impulse response that extends forever, although they may decay. IIR filters may be simpler to implement that FIR filters but may also exhibit some undesirable characteristics like instability. Analog filters are technically IIR filters; for example, capacitor-based filters have an exponential impulse response that in theory never actually reaches zero. Digital FIR filters might be preferred because they can be designed to be **linear phase**, meaning that the delay is equal at all frequencies, preserving the original phase relationships of the signal sinusoidal components.

Physical modeling provides the ability to go beyond theoretical digital filters and mimic the actual physical behavior of analog equipment, ultimately offering the option of working exclusively with digital audio for recording and processing audio. Refinements in digital signal processing (DSP) techniques have yielded convincing software simulations of the hardware devices that have long been the mainstay of analog recording. The physics of analog circuit behavior including real-world component behavior can be modeled in software by making careful measurements that determine which aspects are necessary to recreate the sound imparted by the hardware device. As processor speeds have accelerated, more computation can be accomplished in the time between samples, allowing more refined software models.

Some digital recording systems depend on external DSP hardware; others do all the calculations on the host computer's central processing unit (CPU). Early approaches to simulating analog circuits involved mimicking the sound of the device as a unit; more recent attempts have created physical models of the hardware that simulate the actual circuitry and its physical behavior. These programs are available as add-on software blocks known as **plug-ins**. Plug-ins are desirable in part because one block of code can be used to run several instances of the simulated device, whereas several expensive physical devices would be required to do the same job in a traditional analog studio.

In addition to processor simulations, software instrument simulations are gaining in popularity. Sampled recordings of real instruments have been available for years, forming the basis for many hardware synthesizers. Now it is possible to create software models of physical instrument behavior than can be run in real time to produce sound output directly from control information provided by a mechanical controller. Using pressure transducers and switches, the physical gestures made by a horn player can be fed to a physical model of a horn to produce a very convincing simulation. As our knowledge of the physics of musical instruments is refined, the models of their behavior come closer to producing the same sound as the real instrument.

Digital audio holds the promise of further developments we can barely anticipate. The combination of sophisticated mathematics and fast computers will allow more convincing simulation of analog circuit behavior as well as purely digital techniques that cannot be duplicated in the continuous analog domain. Understanding how digital audio works is important for engineers who simply wish to use the new software systems, as well as for those who wish to further its development.

DIGITAL FILE DISTRIBUTION

Recordings have traditionally been distributed on a physical medium, which gave recording owners something tangible to sell. From Edison's cylinders to 78 rpm phonograph records, LPs, and CDs, there was always a physical medium – until the mp3. The physical medium is being replaced by instant communication nearly anywhere in the world. The compact disc (CD) has been very successful and may ultimately prove to be the only digital medium to remain in use for any length of time. The digital audio tape (DAT) lasted perhaps a decade, mainly as a tool for professionals. DAT sound quality was acceptable, but its mechanical reliability was not – hardly surprising, given the complicated technology involved. A short look at these two media will quickly demonstrate the ingenuity involved.

DAT was originally envisioned as a digital replacement for the compact cassette. It proved too complicated for the consumer but was adopted by audio professionals for two-track recording. DATs are tiny videotape-like cassettes loaded with high-coercivity tape. The tape moves slowly because the heads spin on a cylinder that rotates across the tape at an angle, writing diagonally across the tape rather than along its length. In order to read the tracks, servomechanisms must locate the tracks on the tape and slow or speed the tape to maintain alignment with the written tracks. Location is determined by reading tone bursts recorded alongside the encoded audio signal. The mechanical and electronic control systems were extremely complicated, and careful use and maintenance was required.

Compact discs, once written, are quite impervious to casual damage. The compact disc is an optical medium that uses laser light to read information encoded as edges of tiny depressions in a reflective surface. The trick is that the pit is only 0.6 μm across. They are read by bouncing laser light off the surface and reading the interference pattern of the reflected light. Again, servo systems are required to maintain the alignment of the spinning disc with a movable laser light source and detector. The lack of physical contact required to read the data and the strong polycarbonate plastic used to make CDs contributes to their long life.

Both DAT and CD systems are imperfect as data storage devices. To play back continuously, the data must be retrieved, checked for errors, corrected if possible, and output without a pause. The nature of both magnetic and optical recording media causes a high error rate for raw data. To find and correct errors, an elaborate coding scheme is used. The exact coding scheme differs, but using a special branch of mathematics known as Galois fields, it is possible to build error correction into the digital data encoding method. Therefore, the data on the media are not simply the raw bits but highly modified code symbols that represent the data with redundancy. The raw data are also scattered throughout the data stream to prevent errors from altering all of any one word. All of this detail is necessary to allow digital media to reliably reproduce the recorded signals.

The need for a medium has been reduced with the connection of the Internet to nearly every home. Still, the speed of the Internet doesn't yet allow the broadcasting of full 24-bit 96 kHz stereo sound files. As the speed of the networks slowly increased, it became possible to send music files if they could be compressed into files that contained much less data. The science of psychoacoustics provided the clue in the masking curves (see the discussion of critical bands in Chapter 4). By eliminating any signal components masked by louder ones, the data density could be reduced to a tenth of where it started. This realization gave rise to the mp3 encoder, which breaks down the signal into frequency bands and determines what we would hear in each band. Only the parts we need, determined from analysis of psychoacoustic masking, are returned to the output. Although this and other data compression techniques work surprisingly well, these sound files are not indistinguishable from the uncompressed original sound in careful tests.

The mp3 is a data compression protocol widely used to transmit music files on the Internet. It uses filter banks to simulate the critical bands of the ear. Each filter band is analyzed and converted to individual spectral lines by a modified discrete cosine transform that uses a psychoacoustic model to determine which components would be audible. The data are further encoded and converted to a serial bit stream. The process reduces the data by a factor ranging from $\frac{1}{5}$ to $\frac{1}{50}$, depending on the output bit rate chosen. The reconstructed signal retains enough information to reproduce the original sound, with quality determined by the amount of compression employed.

Algorithms used to compress or otherwise encode data are known as **codecs** (from *code/decode*). Some codecs are termed **lossless** because they do not remove information in the process. Lossless codecs use statistical information about the signal to reduce the amount of data needed to completely reconstruct it. Audio signals have statistical characteristics that allow polynomial equations to approximate them, leaving a small error that is also transmitted. FLAC (Free Lossless Audio Codec) is one example of a lossless codec. Generally, lossless codecs provide larger files than mp3 and other psychoacoustic codecs. Data compression works well enough to have changed the music business for better or worse. That change was inevitable, however, because the speed of communication is increasing daily, and we will eventually be able send our highest-quality recordings anywhere in the world – and possibly even farther than that.

SUGGESTED READING

Pohlmann, K. C. (2011). *Principles of Digital Audio* (6th ed.). McGraw-Hill. ISBN: 978-0-07-166346-5. Covers all aspects of digital audio; a good overall technical text.

Watkinson, J. (2002). *An Introduction to Digital Audio* (2nd ed.). Focal Press. ISBN: 0-240-51643-5. Another overview of digital audio processes.

Smith, J. O. (2007). *Introduction to Digital Filters: With Audio Applications.* W3K Publishing. ISBN: 978-0-9745607-1-7. A good review of digital filters as used in audio.

Monitoring, Mixing, and Mastering

Contents

Once recordings have been made, the task of creating a version optimized for listeners remains. For stereo recordings, this process may involve equalization and gain changes that allow the material to suit the medium intended for distribution, a process known as **mastering**. For multichannel recordings, the individual tracks must first be combined, or **mixed**, in order to create the desired output. Each track may be processed with spectral and/ or dynamic range processing in potentially complex combinations that yield a compelling final mix. In either case, we make decisions about the processing we use by listening to the results. The playback system is a crucial element in the decision making process, as it can affect the perceived sound and therefore our mixing decisions. **Monitoring** involves not only the loudspeakers but also the acoustical environment in which they operate – the listening room. Acoustical problems inherent in the loudspeaker–room interaction cause us to make processing decisions that also reflect the room behavior and are then incorporated into the mix. When such a mix is heard in a different room, it will not sound as we expect.

The Science of Sound Recording
ISBN 978-0-240-82154-2

MONITORING

Monitoring should deliver an unaltered representation of the signals we recorded. This intent is complicated by the interaction of a loudspeaker with the listening room acoustics. We encounter the same issues in the recording studio, where the room acoustics influence the placement of instruments and microphones. Though headphone monitoring eliminates the room contribution, headphones produce a different perception of the spatial arrangement of the sonic elements. Whereas loudspeakers produce a sound that we perceive as originating in space, headphones create a panorama that seems to be located inside our heads. This difference is due to the absence of the directional cues we expect in a normal listening experience. The nervous system creates expectations based on our experience, and if our monitoring lacks or alters these cues, the auditory system fights to establish a coherent image of the sound in space. Headphones can be useful in listening to isolated elements of a recording or mix, as they enhance our ability to hear low-level details that can be masked by room reverberation when using loudspeakers. They are not adequate for judging the overall quality of a mix intended to be heard on loudspeaker playback systems.

The design and placement of loudspeakers are important aspects of studio design. Although the technical performance of the loudspeakers is usually considered, their placement and the resulting acoustical interactions are less readily understood. Because a poorly designed room changes the perceived sound from the best loudspeaker, we need to understand how the acoustical environment behaves when driven by the speakers. This behavior will change with the placement of the loudspeakers and their location relative to reflective and absorptive surfaces. The design of the loudspeaker itself is also a consideration because different designs radiate sound in different patterns in a frequency-dependent way. Analyzing the loudspeaker–room interaction helps us eliminate conditions that alter the sound we hear at the listening position.

LOUDSPEAKER DRIVERS

The electromechanical transducers that convert an electrical signal into sound are called **drivers**. To produce sound from an electronic signal, we rely on the same principles used in microphones. Loudspeakers convert electrical energy into sound by forcing movement in a surface that is in contact with the air, the inverse of the action of a microphone. The loudspeaker is optimized for energy transfer in the opposite direction from the microphone, and power transfer is a prime consideration. In the microphone, we are interested in measuring only the voltage; both current and voltage are required to produce the power necessary to move the considerable mass of a speaker cone. Both inductive and capacitive (electrostatic) systems may be used to construct loudspeaker transducers. As with microphones, the dynamic induction-based system is most common.

Dynamic loudspeakers use the principle of induction to drive a coil, known as a **voice coil**, situated within a strong magnetic field. The signal is applied to the coil, forcing it to move as the current changes. In a typical dynamic driver, the coil is attached to a cone that moves like a piston in response to the electrical signal. Because we are applying a lot of power to the system, we have to consider the heat produced so that the coil doesn't deform or melt. The excursion of the diaphragm should be linear within the limits of the geometry of the driver.

Figure 10-1 shows two dynamic drivers: the cone driver and the compression driver. A **cone driver** uses a cone-shaped surface to move air. The cone can be made of paper, plastic, or metal. A large-diameter cone is necessary to move adequate air; however, the size of the cone surface leads to deviation from the ideal piston-like movement at higher frequencies, a phenomenon known as "break-up." As the driving frequency increases, the cone surface begins to exhibit modes of vibration that cause distortion, which limits the ability of the cone driver to produce the full range of audible sound frequencies.

A second type of dynamic driver is the compression driver. The **compression driver** uses the same induction principle as the cone driver; however, the moving diaphragm compresses air that is fed into a phasing plug that focuses the moving air into the throat of a horn that provides

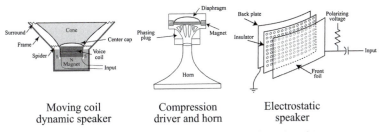

| Moving coil dynamic speaker | Compression driver and horn | Electrostatic speaker |

Figure 10-1 Dynamic and electrostatic loudspeaker drivers.

acoustic impedance matching to the surrounding air. Compression drivers are efficient because they translate small diaphragm movements into large particle velocities that are then focused by the horn. Compression drivers are often used for midrange frequencies – 200 to 2000 Hz – providing low distortion and high sound pressures.

A type of dynamic driver often used for the upper frequency range is the **dome radiator**. The diaphragm is a convex dome suspended and driven by its edge. The convex shape allows the dome to radiate sound with minimal phase distortion. Because the dome is edge-driven, the convex shape guarantees that the dome produces a coherent wave front by placing the center of the dome, where movement is delayed by the propagation time from the edge, closer to the listener. A similar radiator is the **ring radiator**, where the moving surface is a ring rather than a dome. The ring radiator uses diffraction to focus the output sound.

Another driver type is the **ribbon driver**. The ribbon driver is similar in design to the ribbon microphone; however, the ribbon is driven by an applied current that causes it to move in a strong magnetic field. Full-range ribbon speakers are not efficient at low frequencies but are well suited to tweeter use. In one application, a folded ribbon is suspended in a strong magnetic field that causes the ribbon to move when current flows though it. The ribbon squeezes the air between folds as it vibrates, propelling the air outward in both front and rear directions. The squeezing action accelerates the air beyond the velocity of the ribbon itself. The very low mass allows ribbon drivers to operate at frequencies up to 50 kHz or more.

Electrostatic drivers work on a different principle from dynamic drivers, relying on electric attraction and repulsion to move two foil surfaces. Similar to the capacitor microphone, the electrostatic speaker requires a polarizing voltage between the front and back foil. Adding a signal voltage to the polarizing voltage causes the foils to move in proportion to the signal. Because both front and rear foils move, the driver radiates as a dipole, exciting the room behind the loudspeaker as well as in front. This effect must be considered when placing electrostatic speakers. Electrostatic drivers must be large in order to produce significant sound pressures at long wavelengths.

SPEAKER CABINETS

A raw loudspeaker is a dipole; as the cone moves outward in the front, producing a compression wave, it moves inward in the rear, creating a wave of rarefaction. Along the side edge, no sound is radiated. This arrangement is inefficient at moving air; loudspeakers are mounted in cabinets to improve their efficiency. The simplest mounting would be flush with the wall surface. In this situation, called an **infinite baffle**, the speaker cone movement outward produces a compression wave radiating outward while the inward motion of the rear of the cone produces a rarefaction that is not in communication with the listening room. This design requires the loudspeaker to be permanently mounted in the wall – not a practical requirement. By enclosing the loudspeaker in a sealed box, we can approximate a more ideal monopole radiator in which the room is excited only by the radiation from the front of the loudspeaker and is therefore of the same polarity, compression or rarefaction, throughout the room. Although these enclosed speaker cabinets are often referred to as "infinite baffles," they do not behave exactly as an infinite baffle because they also radiate low frequencies around the cabinet to the rear in the same polarity as at the front. Figure 10-2 shows the basic types of cabinets.

The matching of the drivers to the cabinet also depends on acoustical considerations. The cabinet acts like a small room, reflecting internal sound waves that interfere with the driver and cause the cabinet to resonate. The air inside the cabinet can form standing waves, just as we find in a room, so damping materials are placed inside the cabinet to reduce internal interference. The cabinet requires stiff and heavy cabinet walls and internal damping materials to minimize resonance. Each speaker driver has a characteristic resonant frequency determined by the mass and compliance of the device. The compliance of the air in the cabinet combines with the mechanical resonance of the driver to produce a resonance for the total

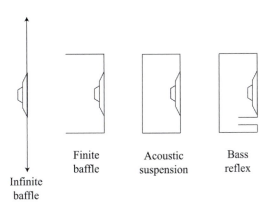

Infinite
baffle

Finite
baffle

Acoustic
suspension

Bass
reflex

Figure 10-2 Speaker cabinet types compared.

system. Below the resonant frequency, the acoustical output drops. Adding ports allows the resonance to shift to lower frequencies, extending the low-frequency performance of the speaker and increasing efficiency.

LOUDSPEAKER–ROOM INTERACTIONS

A loudspeaker placed in a room radiates sound waves outward in directions that depend on the design of the speaker(s) and cabinet. The waves reflect off the room surfaces and begin to interact as direct sound waves encounter reflected sound waves and produce patterns of cancelation and reinforcement. These reflections are perceived initially as discrete echoes and later as rich reverberation. The character of these reflections depends on which frequencies cancel and which reinforce, both factors determined by the wavelengths of signal components and the distances that they travel before reflecting. Because the directional characteristics of the loudspeaker determine how the room is driven, we will consider the common speaker designs and their patterns of radiation.

Speaker Placement

It might seem like the ideal monitoring situation would be for sound produced by the loudspeaker to reach our ears and go nowhere else, as this would remove the room response from consideration. Two methods of accomplishing that result tell a different story: headphones create a different sense of spatial placement and anechoic rooms sound unnatural and unpleasant. We are used to hearing sounds in a natural acoustical environment, so our listening room should be a natural acoustical environment. The trouble results from the natural behavior of the average room. By examining the speaker's radiation pattern and the acoustical behavior of the room, we can minimize the problematic room effects while preserving the sound quality originally delivered by the loudspeakers.

Speaker Radiation Patterns

The radiation pattern from loudspeakers and their cabinets is quite complicated. An ideal sound radiator would be a point source small enough to not have dimensions that could affect the radiation of the sound wave. In reality, we have rather large drivers because they need to move a significant amount of air in order to deliver sound in a relatively large space. This design results in complex patterns of radiation when the entire surface of a drives radiates sound waves that can interfere with each other. At low frequencies at which the wavelength is long relative to the moving driver, sounds radiate more or less omnidirectionally. As we reach higher frequencies, the dimensions of the drivers reach the wavelengths of the signal, and we begin to get interference patterns that narrow and split the radiation into lobes (Figure 10-3).

Because few drivers are capable of undistorted output across the bandwidth of audio signals, most loudspeakers consist of more than one driver. This feature is accomplished by filtering the signal with equalizers called **crossovers**, separating the signal delivered to each driver into frequency ranges compatible with the driver's frequency range. As with all analog filters, there are changes to the phases of signal components, complicating the design of the circuit. When using externally powered loudspeakers, the crossover splits the signal after the power amplifier, requiring high-power components. Most loudspeakers now use internal power amps that allow internal crossovers that act on the line-level signal before the amplifiers that are matched to the driver impedance and range.

In multidriver speakers, there is interference between the drivers when frequencies overlap. Because the crossover filters have finite slopes and must overlap in frequency to allow full-range output, some frequencies near the crossover frequencies are emitted from more than one driver, resulting in interference. Another source of interference is the diffraction that occurs at sharp edges of cabinets. As the sound wave propagating outward from the driver reaches the edge of the front of the cabinet, it reaches an area of lower acoustical impedance as it bends around the corner. This impedance mismatch generates a reflection that radiates back toward the driver and

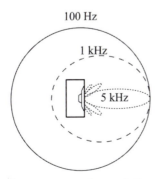

Figure 10-3 As frequency increases, speaker radiation pattern narrows from an omnidirectional to a beamy multilobe shape.

interferes with the wave then being produced. Many cabinets have beveled or rounded edges to reduce this form of interference.

Listening Room Response

As we see in Figure 10-3, the radiation pattern from a loudspeaker is quite complex at higher frequencies. As the pattern grows more complex, so does the interaction with the reflective surfaces of the room. leading to a situation in which the sound perceived varies with listener position. Any attempt to change the perceived sound balance with electronic equalization is therefore futile, as the corrective curve will be different for every point in the room. Problem reflections are best controlled by acoustical treatment. A wide range of absorbing and diffusing products are available, but their application requires knowledge of the nature of the problem.

Effective treatment of acoustical problems depends on understanding what auditory cues are important to preserve and how best to limit undesirable interactions while preserving the desired room response. We have two competing requirements: one that we deliver the spatial cues present in the recorded sound and another that we allow the room to contribute some reverberation, lest we find ourselves in an unpleasant anechoic environment. These criteria must both be met when choosing and treating the listening room.

As we see in Figure 10-4, the reflective surfaces of the listening room create a series of delayed images of the loudspeaker sound called **early reflections**. The time between these discrete echoes gives us information

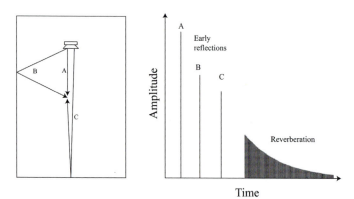

Figure 10-4 Early reflections from room surfaces interfere with the playback sound that already contains early reflections in the recording.

about the size and shape of a room. These reflections are already present in recordings made in rooms. The combination of two sets of early reflections creates confusion that can distract the listener. A goal of acoustical treatment is to allow the recorded signals to mask the contribution of the room in the time period just after a sound is produced. There are several common control room designs that attempt to create this situation, though in different ways. Some separate the front and back of the room and treat each separately; others try to minimize the contribution of the entire room. The complexity of the issue is reflected in the number of attempts at solution.

Several systems for treating the control room have been employed in an attempt to provide a neutral acoustical environment. The **Live-End-Dead-End (LEDE)** system mounts the loudspeakers in the front wall to eliminate side- and rear-oriented radiation while treating the front of the room with absorbers. The rear of the room is treated with broadband diffusers to maintain a reverberant space that does not interfere with the recorded early reflections. An alternative approach is the **nonenvironment** system. This system also mounts the loudspeakers in the front wall; however, the entire room is designed to present similar acoustics. The rear and side walls and ceiling are absorbent, while the floor and front wall are hard reflective surfaces. This design allows sounds not originating from the loudspeaker, like speech, to have a normal reverberant quality, while early reflections from the loudspeaker are damped. The necessity to provide absorption down to low frequencies makes this approach diffi-cult, as it is increasingly difficult to remove sound energy as the frequency drops below 100 Hz.

Many studios have adopted the so-called **near-field monitor** as another method of reducing the contribution of the room. (These monitors are more properly called **close-field** monitors to avoid the technical definitions of the near field that involve the geometry of the radiating sound waves.) The close field is operationally defined as the area in which the direct sound from the loudspeaker masks the reverberant field and room resonances. Close-field monitors are placed close to the engineer at the mix position in the hope that the proximity to the listener will allow the direct sound to mask any early reflections created by off-axis radiation from the loud-speakers. Such close monitoring has limitations, such as reflections from the mixing console when the monitors are placed on the meter bridge, as they often are. Too close to the speaker, the sound from the separate drivers may change as the listener's head moves, changing the sound perceived. The growth of small studios has increased the demand for close-field monitors.

Figure 10-5 A typical treated studio room. Front and side walls are absorbent with movable diffusers; the floor is reflective and the ceiling absorbent.

Figure 10-5 shows a studio treated with absorbers and diffusers. The diffusers can be covered by sliding absorbing panels to allow some variation of the room acoustics. Corners are bass traps designed to remove the low-frequency energy that tends to increase in corners, where reflected waves reinforce with incident waves, especially at long wavelengths.

MIXING

The combination and modification of recorded tracks into a final form is known as **mixing**. The mixing process takes advantage of electronic or software processing tools that affect the amplitude, apparent spatial placement, dynamic range, and spectral content of the recorded signals. These basic processes are each relatively simple conceptually; however, the combination of these processes allows great creativity in the assembly of the final mix of sounds. The goal of mixing is to provide a sonic experience that maximizes the impact of the presentation regardless of whether it is music or other sounds. In order to take full advantage of the possibilities, we need an understanding of what is possible using the tools at our disposal.

Panning

The basic mixing console provides control over the signal level and the placement from left to right in the panorama for each input signal. A fader is used to set the amplitude of each signal, and a pan potentiometer is used to place the phantom image of the signal somewhere between the left and right loudspeaker. This approach provides only half of the spatial information we normally use to localize a sound source; it is missing any time differences that would accompany natural sounds. The intensity information used in panning creates a less than perfect apparent image placement; however, it is used so frequently that it has been accepted as a reasonable approximation.

The pan control consists of two matched potentiometers connected in opposite polarities so that as one increases, the other decreases. The outputs are routed to the left and right speakers so that by turning the potentiometer, the signal appears to move. As a mono signal is panned from left to right, the center position is potentially going to be 6 dB louder than when the signal is panned to one side or the other. This effect is often compensated for by what is called a **pan law**. The pan law determines what happens as the signal is panned through the center of the image; it may be a user option in software mixers but is fixed in hardware mixers. Because the center image intensity depends on acoustic mixing to combine the outputs of the two speakers, in practice less than 6 dB is usually perceived due to room interference. A common pan law gives a 3 dB reduction at center relative to either extreme.

Compression

The most powerful and commonly used processing techniques applied during mixing are dynamic range compression and equalization. Each has variants that sound different and give the engineer a large palette with which to work. Understanding the nuances of each available tool makes the job of mixing more fruitful and more enjoyable. Probably the biggest difference between "beginning" mixes and what we hear on commercial releases is the artistic use of dynamic range compression. Without using compression, it is often impossible to make a full-sounding mix with clarity and punch. Compression can also ruin a mix if it's overused, so compressors are an important tool to master. Although compression is easy to understand conceptually, it is touchy and takes plenty of experimentation and careful listening to master in all its manifestations.

The basic idea of compression is to reduce the **dynamic range** – that is, the difference between the loudest and softest elements of a sound. The compressor does this by changing the gain of an amplifier based on the amplitude of the input signal. When the signal crosses the threshold level, the linear gain of the amplifier is reduced to a lower gain, thus reducing the output amplitude. If the output gain is then increased (using make-up gain) so that the overall sound is as loud as before, the softer elements are amplified and become louder relative to the louder elements. The exact method for accomplishing this effect electronically creates sonic differences between compressors.

Some compressors make dramatic alterations in the sound character; others seem much more transparent. Each type has its uses, but there is significant variability in compressors due to the exact type of variable-gain element and signal-level measurement circuitry employed. Gain elements include voltage-controlled amplifiers (VCAs), FETs, photo-optical elements, and variable-mu vacuum tubes. Each imparts a different sound characteristic; often a compressor is selected for a particular track based on its sonic signature. The amplitude measuring circuitry can also contribute to the sound: RMS detectors more closely reflect our sensory loudness perception than do average detectors.

The time course of compression onset and release also contributes to the final sound quality. A fast-attack compressor reduces the onset transients of sounds; longer attack times leave the transient alone and compress the later sustained sound more. Release times that are too short make the gain change more obvious (this is called **pumping**). Different compressors have different controls available: some use automatic attack and release, some have no threshold controls, some have continuously variable controls, and some fixed offer steps. Understanding the nuances of the behavior of different types of compressors takes time but is well worth pursuing.

In addition to the standard methods of employing compression, many methods exist to creatively connect compressors for specific applications. Compressors with **sidechain** or **key** inputs allow signals other than the one being compressed to modify the behavior of the device. For example, lead vocals may be sent to a key input on a rhythm instrument to allow the vocal to decrease the gain of the instrument when there is singing present. This process is known as **ducking**, as the vocal causes the instrument being compressed to duck down in level to make more room for the vocal in the mix. Compressors can be used on several signals at once by bussing them together and routing the bus to an aux return with a compressor. Many compressors allow an external device to process the signal that is used to derive the control signal separately from the sound being compressed through a **sidechain insert**. For example, this practice allows the compression to be controlled differently for different frequencies in the signal by patching an equalizer into the side chain. If the equalizer is a high pass filter, the bass can be prevented from causing the gain to change, eliminating the pumping that results when the bass modulates the gain of higher frequencies. By selectively boosting a band of higher frequencies in the side chain, the compressor can be made to reduce vocal sibilants in a process known as **de-essing**.

Another useful compression technique is **serial compression**, in which two or more compressors are used in series. This method allows each compressor to do a little compression and avoids overdriving the sidechain electronics, as might occur with heavy compression using a single device. Another useful technique is **parallel compression**, in which a signal is delivered to the mix uncompressed and fed to a compressor whose output is mixed with the uncompressed signal. This approach allows the original dynamics of a track to be preserved while adding some compressed signal to "fatten it up." With analog devices, these techniques create no timing problems, but on **digital audio workstations** (DAWs), the plug-in delays must be compensated for to avoid the comb filtering that results from unequal time delays.

Equalization

Like compressors, there are many types of equalizers available. For sweeping changes in spectral content, shelving filters allow boosting or cutting all frequencies above or below a selected frequency. Parametric filters allow surgical precision in boosting or cutting adjustable regions of frequencies. Graphic equalizers create up to 31 separate overlapping narrow-band filters that give control over the entire audible frequency range. Combinations of these filter types may be used as well. Again, it takes some experience to know which type of filter best suits a given mix, but some generalizations may be of use: cutting is usually preferable to boosting, as it does not increase the noise level and small changes are often preferable to large ones.

Equalizers are available in a number of configurations from fully parametric to graphic. In computer audio programs, common plug-in equalizer programs include high- and low-pass, shelving, and parametric bands that can be used independently or combined. Each filter type best addresses different demands. Shelving filters allow great alterations in the lowest and highest frequencies and are commonly available as tone controls on stereos for that reason. Fully parametric filters work best to eliminate or boost small frequency bands, as they can be carefully tailored to change just what we need without affecting other frequencies. Most often, it is best to use small boosts and cuts together rather than large boosts or cuts alone. Changes of more than a few decibels are seldom necessary if proper balancing of boosts and cuts is employed.

Every mix is different, but some common problems can be reduced using generalized equalization approaches. Every element of a mix has a characteristic spectral content, but most instruments produce strong energy in the midrange, where our ear is most sensitive. This effect leads to complex masking issues as sounds come and go in different parts of a song. Removing just a little midrange (around 1–2 kHz) from most elements will allow the main focus sound, often vocals, to be accentuated by leaving their midrange content unfiltered. There are also areas in the low frequencies where a buildup occurs, often around 300 Hz. Similar gentle filtering can clear the low end of congestion and make the sounds more discrete or tight-sounding without sounding thin. Only a couple of decibels of reduction is usually sufficient and does not greatly change the character of the individual sounds. Slightly boosting sounds in the 2–6 kHz range can make them cut through a mix without substantial gain increases. Many small changes can make a big difference in a mix.

Time Delay Effects

Before digital audio made time delay easy, effects that require delays relied on tape recorders to create delayed copies of a signal. Taking the output from the playback head delayed the signal by the tape speed divided by the distance between the record and play heads. If the tape machine had vari-speed (variable tape speed), the delay time could be adjusted. Routing some of the delayed signal back to the input created a repeated echo. **Flanging** involved running two tape recorders playing the same program and slowing one, then the other, by rubbing a reel flange with a hand while the two signals were mixed. The development of digital audio made time delay simple, and the old effects are easily simulated using digital delays.

The simplest delay effect is a single delay, which can be used to increase the apparent size of a sound by slightly delaying it and adding it back to the original sound. Panning different short delays to the left and right (less than 25–30 ms) creates a sense of space without also creating the perception of a discrete echo. Longer delays are heard as discrete echoes and often used on vocals. Vocal delays can be synchronized with the tempo of a song, which

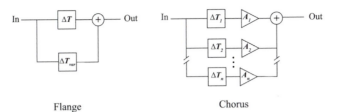

Flange   Chorus

Figure 10-6 Flanging and chorus effects use delay to create comb filtering that generates a distinctive spatial effect.

makes echoes coincide with the rhythmic pattern, reinforcing their impact. Dividing 60,000 by the tempo in beats per minute gives a quarter-note's duration in milliseconds. By setting the delay time to a rhythmic value, quarter-note, eighth-note, dotted-quarter, or other related value, the delay will sound like a natural part of the sound. Unrelated delay times produce a repeated echo that is out of time with the rhythm.

Complex sounds are created when delayed signals are recombined with the original. Flanging is produced by summing changing delays that produce **comb filtering** as various frequencies cancel and reinforce. This technique produces a rich, full sound effect. **Chorus** is a similar effect. It involves multiple separate delays, each with a variable delay and amplitude envelope, that are combined to produce an effect simulating the sound of many singers.

Reverberation simulates the reverberant behavior of a space. For many years, it was accomplished using a loudspeaker and microphone in a real room or chamber. Tensioned steel plates were also used to create simulated reverberation. These methods are now modeled using digital signal processing to emulate the behavior of the mechanical devices whose sound became associated with popular music. It is also now possible to make very convincing simulations of real halls and rooms directly. Digital reverberation offers the ability to place recordings made in dry rooms into any sonic environment we desire.

Reverberation as complex as that of a real room is a challenge to simulate. Modeling the acoustics of a room demands considerable computing power, especially if it must be done in real time. Each separate path from source to listener can be characterized by a delay and filter with variable gain to simulate the absorption in that path, but considering how many paths are required to create a natural sounding reverberation, this is too computationally expensive. Another approach to creating reverberation makes use of the impulse response of a space. Using convolution, a signal is combined with the impulse response to simulate how the sound would reverberate in the space. Convolution reverbs allow new reverberant spaces to be sampled with impulse response recordings but may lack some of the parameter adjustments available with other methods of simulating reverberation.

Automation

Automation, the ability of the mixing system to remember and repeatedly execute the moves we make on the controls, is now accepted as an integral part of mixing. It was not always so: early automation systems could only play back fader moves using complex outboard add-on hardware. Before that, mixes were often team efforts and sometimes had to be edited together from multiple takes in which each produced a different part of the song satisfactorily. Modern systems can automate every change possible, including controls inside the plug-ins. It is certainly easy to be sucked into a vortex of automated changes that micromanage a mix. Although level and mute automation can be very useful and an occasional plug-in may benefit from automation in certain cases, this feature should not be allowed to take over the job of mixing. The benefit of automation comes from handling certain problems, but overuse can create new ones if we're not careful.

One common use of gain automation comes with vocal tracks on which the levels are wildly varying, which frequently happens in songs with wide dynamics. If we rely on a compressor, or even a chain of compressors, we are likely to find that though the mix sounds good in the loud parts, the compressor doesn't produce what we want on the quiet parts, or vice versa. By balancing the levels of the vocal track with automation, the compressor will then act similarly on all sections of the vocal track. This technique may apply to any track with wide dynamics that we wish to tame. It is like having an engineer ride the faders each time the mix is played. Be sure your compressor is postfader, however, or this approach will not produce the desired effect.

Breakpoint envelope automation on a DAW allows one to sculpt the gain of a track while observing the waveform, a feature not available when using traditional fader automation. This automation makes small, accurate changes tightly linked to the signal simple; however, it can also take time to do these adjustments. These envelopes can be copied and pasted to other tracks, so backing vocals can be jointly processed, for example. Fades can be done this way, as well. These changes can also be done using the virtual fader and mouse in real time as well as by using the graphical interface. In addition to amplitude, most parameters available can be automated in this way.

Automation of mutes is another useful technique. Frequently, a microphone is recording throughout a song when the instrument we are recording contributes to the sound only infrequently. Gates sometimes control this situation adequately, but if the threshold cannot clearly discriminate between leakage and the desired sound, a gate may not give us the right performance. In such cases, we may automate mutes to allow us the precise control of when we hear the sound and when we do not. This method can be time-consuming, but if done carefully it can clean up unwanted leakage without altering the desired elements of sound. Panning may also be automated – an effect that can easily be overused.

MASTERING

Mastering has traditionally meant assembling a collection of tracks that becomes a work in itself with consistent levels and tonality throughout. Mastering was a necessary step when recordings were released on physical media. Vinyl records had strict limitations in terms of amplitude and frequency content that required close monitoring, and the final product was cut on a lathe in real time by a skilled engineer. Cassettes and later CDs had their own limitations that required audio program adjustments to get the most from those media. As distribution moves away from albums on physical media, the concept of mastering takes on new expectations, often making individual mixes louder to compete with other releases. In order to know what a master recording will sound like, the decisions need to be made in a controlled acoustical space. Often, changes of a fraction of a decibel at particular frequencies or a few decibels of limiting will be quite audible in a mastering environment. Through experience, a mastering engineer knows how what he or she hears in the studio will sound on a wide range of systems listeners will use.

Mastering is often performed using analog equipment. Most recording projects now involve digital recording, and software exists to perform the mastering process, but frequently the digital mixes are played back in analog and rerecorded through a high-quality A/D converter after analog processing. Digital mixes are also sometimes sample rate converted by the same procedure. Some mastering engineers still prefer their familiar analog processing systems to experimentation with new digital versions. Programs for mastering do allow easy manipulation of track timing, order, levels, and metering. Many mastering engineers use these along with analog processing to get the best possible final product.

The goal of mastering is to provide the best sound possible for the intended release medium. For some musicians and producers, the desire is for the final recording to sound as if it were being played on the radio. Most music listeners are familiar with the characteristic compressed sound produced by radio broadcasters. That sound is influenced by the stations' desire to be the loudest signal as the dial is spun from station to station – a result of the psychoacoustic preference for the louder of two sounds. It is achieved by heavy compression and limiting in the air chain, which keeps the antenna radiating near the maximum power allowed all the time. This series of specialized compressors and limiters is designed to operate on recordings with "average" dynamics, converting it to the "radio sound." If the recording is mastered to sound like the standard air chain, playing it through an air chain will not have the desired effect. When the material is already so heavily compressed, the additional compression and limiting ironically makes the sound seem smaller.

Because many of today's recordings are distributed by digital down-loading, the effect of data compression should be a consideration in the mastering process. Lossy codecs like the mp3 result in alterations of the signal that may lead to distortion that can be minimized in the process of mastering. High frequencies and transients are particularly challenging for codecs to preserve. Especially at low bit rates, the quality of the compressed signals is significantly reduced. If recordings are intended for low-bit-rate encoding, limiting the high-frequency content can reduce the perceived distortion generated in the compression process.

In an era of reduced budgets, many recording engineers are asked to master their projects. There are good reasons why mastering is ideally a separate step. Even well-designed control rooms have acoustical signatures, and these are incorporated into the mixed sound. Mastering studios are built to be as neutral-sounding as possible and, with fewer constraints than control rooms, often come much closer to the ideal. The mastering studio needs to create the ideal sound balance for a smaller audience than the control room. The room is optimized for careful listening and tailored to that one task. Though there is less necessary gear in a mastering studio, the quality of each piece is paramount. The electronic equipment is optimized for accuracy and repeatability, giving tighter control of the dynamics pro-cessing and equalization that make up much of the mastering process.

The experience of a mastering engineer should not be ignored. They have heard project after project and have acquired knowledge of how each has translated in the outside world. That understanding is brought to each new project. Although it is certainly possible to check a master on a car stereo and an iPod, it is time-consuming and distracts from the process. It takes time to produce a suitable master if each one must be checked for balance on the whole range of possible playback systems. Of course, this might be the first step taken by a prospective mastering engineer.

One advantage of professional mastering is the ability to hear the mixes on an excellent playback system. Mastering studios are able to spend much of their budget on loudspeakers and amplifiers, providing a sonic microscope that reveals the slightest details in recordings. High-quality mastering processors – like compressors, limiters, and equalizers – are capable of tiny but audible changes that ordinary processors can't provide. By hearing our mixes in that environment, we learn what may be deficient in our own monitoring systems. Observing what is done to "fix" our mixes helps the next mixing job to come closer to the ideal. Experimenting with mastering can ultimately be time well spent, but paying for professional mastering is a learning experience not to be overlooked.

SUGGESTED READING

Katz, B. (2007). *Mastering Audio: The Art and the Science* (2nd ed.). Focal Press. ISBN: 978–0240808376. Covers mastering techniques in detail.

Newell, P. (2003). *Recording Studio Design*. Focal Press. ISBN: 978–0240519173. Covers studio acoustics and construction.

Izhaki, R. (2008). *Mixing Audio: Concepts, Practices, and Tools*. Focal Press. ISBN: 978–0240520681. Mixing techniques explained.

APPENDIX 1

Here are some examples of calculations involving decibels. Decibels are ratio measures that depend on standard reference values against which a variable amplitude is compared.

Example 1: We are connecting a device with a nominal output level of -10 *dBV* to a device with a specified input level of $+4$ *dBu*. When a signal is at the nominal amplitude, the device meter reads 0 *VU*. What will the input *VU* meter read when fed from the output?

$$dB = 20 \log\left(\frac{V_{measured}}{V_{reference}}\right)$$

First we convert -10 *dBV* to voltage:

$$-10 \; dB = 20 \log\left(\frac{V}{1.0}\right) = 20 \log V$$

$$-\frac{10}{20} = \log V$$

$$10^{-1/2} = V = \frac{1}{\sqrt{10}} = .316 \; V$$

The output voltage (-10 *dBV*) is 0.316 *V*:

$$20 \log\left(\frac{.316}{.775}\right) = 20 \log(.408) = -7.8 \; dBu$$

Because the input expects an input of $+4$ *dBu*, the meter will read -7.8–4 *dB*, or -11.8 *VU*.

Example 2: How much peak-to-peak voltage is necessary to allow a $+4$ *dBu* RMS sinusoidal signal to pass through a circuit undistorted?

$$+4 \; dBu = 20 \log\left(\frac{x}{.775}\right)$$

$$\frac{4}{20} = \log\left(\frac{x}{.775}\right)$$

$$10^{0.2} = \left(\frac{x}{.775}\right)$$

$$1.58 \times .775 = 1.23 \; V \; RMS$$

From Figure 1-5, peak-to-peak measures are 2.828 times RMS measures for sine waves:

$$1.23 \times 2.828 = 4.1 \ V \text{ p-p}$$

The power supply needs to provide at least 4.1 volts peak-to-peak to accommodate a +4 *dBu* signal.

For a signal of +24 *dBu* RMS (a good amount of headroom), we would need a peak-to-peak voltage of:

$$24 = 20 \log\left(\frac{x}{.775}\right)$$

$$\frac{24}{10} = \log\left(\frac{x}{.775}\right)$$

$$10^{1.2} \times .775 = 12.3 \ V \text{ RMS}$$

$$12.3 \times 2.828 = 35 \ V \text{ peak-to-peak}$$

This voltage is beyond the limits of op-amps that run on standard $+/-15 \ V$ power supplies.

INDEX

Note: Page numbers followed by *f* indicate figures.

Included here for your reference is Chapter 46 (Test and Measurement, written by Pat Brown) from the 4th edition of the Handbook for Sound Engineers.

The 4th edition of this trusted book has been updated to reflect changes in the industry since the publication of the 3rd edition in 2002 – including new technologies like software-based recording systems such as Pro Tools and Sound Forge; digital recording using MP3, wave files, and others; and mobile audio devices such as iPods and MP3 players.

The latest edition incorporates over 40 topics, each covered and written by top professionals for their area in the field, including Glen Ballou on interpretation systems, intercoms, assistive listening, and image projection; Ken Pohlmann on compact discs and DVDs; David Miles Huber on MIDI; Dr. Eugene Patronis on amplifier design and outdoor sound systems; Bill Whitlock on audio transformers and preamplifiers; Pat Brown on funda-mentals and gain structures; Ray Rayburn on virtual systems and digital interfacing; and Dr. Wolfgang Ahnert on computer-aided sound system design and acoustics for concert halls.

ISBN: 978-0-240-80969-4

CHAPTER FORTY-SIX

Test and Measurement

by Pat Brown

Contents

46.1. TEST AND MEASUREMENT

Technological advancements in the last two decades have given us a variety of useful measurement tools, and most manufacturers of these instruments provide specialized training on their use. This chapter will examine some principles of test and measurement that are common to virtually all measurement systems. If the measurer understands the principles of measurement, then most any of the mainstream measurement tools will suffice for the collection and evaluation of data. The most important prerequisite to performing meaningful sound system measurements is that the measurer has a solid understanding of the basics of audio and acoustics. The question "How do I perform a measurement?" can be answered much more easily than "What should I measure?" This chapter will touch on both, but readers will find their measurements skills will relate directly to their understanding of the basic physics of sound and the factors that produce good sound quality. The whole of this book will provide much of the required information.

46.1.1. Why Test?

Sound systems must be tested to assure that all components are functioning properly. The test and measurement process can be subdivided into two major categories-electrical tests and acoustical tests. Electrical testing mainly involves voltage and impedance measurements made at component interfaces. Current can also be measured, but since the setup is inherently more complex it is usually calculated from knowledge of the voltage and impedance using Ohm's Law. Acoustical tests are more complex by nature, but share the same fundamentals as electrical tests in that some time varying quantity (usually pressure) is being measured. The main difference between electrical and acoustical testing is that the interpretation of the latter must deal with the complexities of 3D space, not just amplitude versus time at one point in a circuit. In this chapter we will define a loudspeaker system as a number of components intentionally combined to produce a system that may then be referred to as a loudspeaker. For example, a woofer, dome tweeter, and crossover network are individual components, but can be combined to form a loudspeaker system. Testing usually involves the measurement of systems, although a system might have to be dissected to fully characterize the response of each component.

46.2. ELECTRICAL TESTING

There are numerous electrical tests that can be performed on sound system components in the laboratory. The measurement system must have specifications that exceed the equipment being measured. Field testing need not be as comprehensive and the tests can be performed with less sophisticated instrumentation. The purpose for electrical field testing includes:

1. To determine if all system components are operating properly.
2. To diagnose electrical problems in the system, which are usually manifested by some form of distortion.
3. To establish a proper gain structure.

Electrical measurements can aid greatly in establishing the proper gain structure of the sound system. Electrical test instruments that the author feels are essential to the audio technician include:

- ac voltmeter.
- ac millivoltmeter.
- Oscilloscope.
- Impedance meter.
- Signal generator.
- Polarity test set.

It is important to note that most audio products have on-board metering and/or indicators that may suffice for setting levels, making measurements with stand-alone meters unnecessary. Voltmeters and impedance meters are often only necessary for troubleshooting a nonworking system, or checking the accuracy and calibration of the on-board metering.

There are a number of currently available instruments designed specifically for audio professionals that perform all of the functions listed. These instruments need to have bandwidths that cover the audible spectrum. Many general purpose meters are designed primarily for ac power circuits and do not fit the wide bandwidth requirement.

More information on electrical testing is included in the chapter on gain structure. The remainder of this chapter will be devoted to the acoustical tests that are required to characterize loudspeakers and rooms.

46.3. ACOUSTICAL TESTING

The bulk of acoustical measurement and analysis today is being performed by instrumentation that includes or is controlled by a personal computer. Many excellent systems are available, and the would-be measurer should select the one that best fits their specific needs. As with loudspeakers, there is no clear-cut best choice or one size fits all instrument. Fortunately an understanding of the principles of operating one analyzer can usually be applied to another after a short indoctrination period. Measurement systems are like rental cars—you know what features are there; you just need to find them. In this chapter I will attempt to provide a sufficient overview of the various approaches to allow the reader to investigate and select a tool to meet his or her measurement needs and budget. The acoustical field testing of sound reinforcement systems mainly involves measurements of the sound pressure fluctuations produced by a loudspeaker(s) at various locations in the space. Microphone positions are selected based on the information that is needed. This could be the on-axis position of a loudspeaker for system alignment purposes, or a listener seat for measuring the clarity or intelligibility of the system. Measurements must be made to properly calibrate the system, which can include loudspeaker crossover settings, equalization, and the setting of signal delays. Acoustic waveforms are complex by nature, making them difficult to describe with one number readings for anything other than broadband level.

46.3.1. Sound Level Measurements

Sound level measurements are fundamental to all types of audio work. Unfortunately, the question "How loud is it?" does not have a simple answer. Instruments can easily measure sound pressures, but there are many ways to describe the results in ways relevant to human perception. Sound pressures are usually measured at a discrete listener position. The sound pressure level may be displayed as is, integrated over a time interval, or frequency weighted by an appropriate filter. Fast meter response times produce information about peaks and transients in the program material, while slow response times yield data that correlates better with the perceived loudness and energy content of the sound.

A sound level meter consists of a pressure sensitive microphone, meter movement (or digital display), and some supporting circuitry, Fig. 46-1. It is used to observe the sound pressure on a moment-by-moment basis, with the pressure displayed as a level in decibels. Few sounds will measure the same from one instant to the next. Complex sounds such as speech and music will vary dramatically, making their level difficult to describe without a graph of level versus time, Fig. 46-2. A sound level meter is basically a voltmeter that operates in the acoustic domain.

Figure 46-1 A sound level meter is basically a voltmeter that operates in the acoustic domain. Courtesy Galaxy Audio.

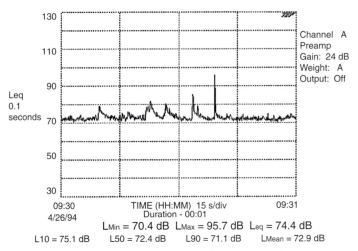

Figure 46-2 A plot of sound level versus time is the most complete way to record the level of an event. Courtesy Gold Line.

Sound pressure measurements are converted into decibels ref. 0.00002 pascals. See Chapter 2, Fundamentals of Audio and Acoustics, for information about the decibel. Twenty micropascals are used as the reference because it is the threshold of pressure sensitivity for humans at midrange frequencies. Such measurements are referred to as *sound pressure level* or L_P (level of sound pressure) measurements, with L_P gaining acceptance among audio professionals because it is easily distinguished from L_W (sound power level) and L_I (sound intensity level) and a number of other L_X metrics used to describe sound levels. Sound pressure level is measured at a single point (the microphone position). Sound power measurements must consider all of the radiated sound from a device and sound intensity measurements must consider the sound power flowing through an area. Sound power and sound intensity measurements are usually performed by acoustical laboratories rather than in the field, so neither is considered in this chapter. All measurements described in this chapter will be measurements of sound pressures expressed as levels in dB ref. 0.00002 Pa.

Sound level measurements must usually be processed for the data to correlate with human perception. Humans do not hear all frequencies with equal sensitivity, and to complicate things further our response is dependent on the level that we are hearing. The well-known Fletcher-Munson curves describe the frequency/level characteristics for an average listener, see Chapter 2. Sound level measurements are passed through weighting filters that make the meter "hear" with a response similar to a human. Each scale correlates with human hearing sensitivity at a different range of levels. For a sound level measurement to be meaningful, the weighting scale that was used must be indicated, in addition to the response time of the meter. Here are some examples of meaningful (if not universally accepted) expressions of sound level:

- The system produced an $L_P = 100$ dBA (slow response) at mix position.
- The peak sound level was $L_A = 115$ dB at my seat.
- The average sound pressure level was 100 dBC at 30 ft.
- The loudspeaker produced a continuous L_P of 100 dB at one meter (assumes no weighting used).
- The equivalent sound level L_{EQ} was 90 dBA at the farthest seat.

Level specifications should be stated clearly enough to allow someone to repeat the test from the description given. Because of the large differences between the weighting scales, it is meaningless to specify a sound level without indicating the scale that was used. An event that produces an $L_P = 115$ dB using a C scale may only measure as an $L_P = 95$ dB using the A scale.

The measurement distance should also be specified (but rarely is). Probably all sound reinforcement systems produce an $L_P = 100$ dB at some distance, but not all do so at the back row of the audience!

L_{pk} is the level of the highest instantaneous peak in the measured time interval. Peaks are of interest because our sound system components must be able to pass them without clipping them. A peak that is clipped produces high levels of harmonic distortion that degrade sound quality. Also, clipping reduces the crest factor of the waveform, causing more heat to be generated in the loudspeaker causing premature failure. Humans are not extremely sensitive to the loudness of peaks because our auditory system integrates energy over time with regard to loudness. We are, unfortunately, susceptible to damage from peaks, so they should not be ignored. Research suggests that it takes the brain about 35 ms to process sound information (frequency-dependent), which means that sound events closer together than this are blended together with regard to loudness. This is why your voice sounds louder in a small, hard room. It is also why the loudness of the vacuum cleaner varies from room to room. Short interval reflections are integrated with the direct sound by the ear/brain system. Most sound level meters have slow and fast settings that change the response time of the meter. The slow setting of most meters indicates the approximate root-mean-square sound level. This is the effective level of the signal, and should correlate well with its perceived loudness.

A survey of audio practitioners on the Syn-Aud-Con e-mail discussion group revealed that most accept an $L_P = 95$ dBA (slow response) as the maximum acceptable sound level of a performance at any listener seat for a broad age group audience. The A weighting is used because it considers the sound level in the portion of the spectrum where humans are most easily annoyed and damaged. The slow response time allows the measurement to ignore short duration peaks in the program. A measurement of this type will not indicate true levels for low-frequency information, but it is normally the mid-frequency levels that are of interest.

There exist a number of ways to quantify sound levels that are measured over time. They include:
- L_{PK}—the maximum instantaneous peak recorded during the span.
- L_{EQ}—the equivalent level (the integrated energy over a specified time interval).
- L_N—where L is the level exceeded N percent of the time.
- L_{DEN}—a special scale that weights the gathered sound levels based on the time of day. DEN stands for day-evening-night.
- **DOSE—*a measure of the total sound exposure.***

A variety of instruments are available to measure sound pressure levels, ranging from the simple sound level meter (SLM) to sophisticated data-logging equipment. SLMs are useful for making quick checks of sound levels. Most have at least an A- and C-weighting scale, and some have octave band filters that allow band-limited measurements. A useful feature on an SLM is an output jack that allows access to the measured data in the form of an ac voltage. Software applications are available that can log the meter's response versus time and display the results in various ways. A plot of sound level versus time is the most complete way to record the level of an event. Fig. 46-2 is such a measurement. Note that a start time and stop time are specified. Such measurements usually provide statistical summaries for the recorded data. An increasing number of venues monitor the levels of performing acts in this manner due to growing concerns over litigation about hearing damage to patrons. SLMs vary dramatically in price, depending on quality and accuracy.

All sound level meters provide accurate indications for relative levels. For absolute level measurements a calibrator must be used to calibrate the measurement system. Many PC-based measurement systems have routines that automate the calibration process. The calibrator is placed on the microphone, Fig. 46-3, and the calibrator level (usually 94 or 114 dB ref. 20 μPa) is entered into a data field. The measurement tool now has a true level to use as a reference for displaying measured data.

Figure 46-3 A calibrator must be fitted with a disc to provide a snug fit to the microphone. Most microphone manufacturers can provide the disc.

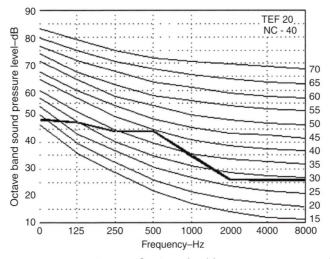

Figure 46-4 A noise criteria specification should accompany a sound system specification.

Noise criteria ratings provide a one-number specification for allowable levels of ambient noise. Sound level measurements are performed in octave bands, and the results are plotted on the chart shown in Fig. 46-4. The NC rating is read on the right vertical axis. Note that the NC curve is frequency-weighted. It permits an increased level of low-frequency noise, but becomes more stringent at higher frequencies. A sound system specification should include an NC rating for the space, since excessive ambient noise will reduce system clarity and require additional acoustic gain. This must be considered when designing the sound system. Instrumentation is available to automate noise criteria measurements.

46.3.1.1. Conclusion

Stated sound level measurements are often so ambiguous as to become meaningless. When stating a sound level, it is important to indicate:
1. The sound pressure level.
2. Any weighting scale used.
3. Meter response time (fast, slow or other).
4. The distance or location at which the measurement was made.
5. The type of program measured (i.e., music, speech, ambient noise).
Some correct examples:
• "The house system produced 90 dBA-Slow in section C for broadband program."

- "The monitor system produced 105 dBA-Slow at the performer's head position for broadband program."
- "The ambient noise with room empty was NC-35 with HVAC running."

In short, if you read the number and have to request clarification then sufficient information has not been given. As you can see, one-number SPL ratings are rarely useful.

All sound technicians should own a sound level meter, and many can justify investment in more elaborate systems that provide statistics on the measured sound levels. From a practical perspective, it is a worthwhile endeavor to train one's self to recognize various sound levels without a meter, if for no other reason than to find an exit in a venue where excessive levels exist.

46.3.2. Detailed Sound Measurements

The response of a loudspeaker or room must be measured with appropriate frequency resolution to be characterized. It is also important for the measurer to understand what the appropriate response should be. If the same criteria were applied to a loudspeaker as to an electronic component such as a mixer, the optimum response would be a flat (minimal variation) magnitude and phase response at all frequencies within the required pass band of the system. In reality, we are usually testing loudspeakers to make sure that they are operating at their fullest potential. While flat magnitude and phase response are a noble objective, the physical reality is that we must often settle for far less in terms of accuracy. Notwithstanding, even with their inherent inaccuracies, many loudspeakers do an outstanding job of delivering speech or music to the audience. Part of the role of the measurer is to determine if the response of the loudspeaker or room is inhibiting the required system performance.

46.3.2.1. Sound Persistence in Enclosed Spaces

Sound system performance is greatly affected by the sound energy persistence in the listening space. One metric that is useful for describing this characteristic is the reverberation time, T_{30}. The T_{30} is the time required for an interrupted steady-state sound source to decay to inaudibility. This will be about 60 dB of decay in most auditoriums with controlled ambient noise floors. The T_{30} designation comes from the practice of measuring 30 dB of decay and then doubling the time interval to get the time required for 60 dB

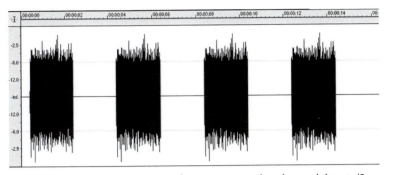

Figure 46-5 Level versus time plot of a one-octave band gated burst (2-second duration).

of decay. A number of methods exist for determining the T_{30}, ranging from simple listening tests to sophisticated analytical methods. Fig. 46-5 shows a simple gated–noise test that can provide sufficient accuracy for designing systems. The bursts for this test can be generated with a WAV editor. Bursts of up to 5 seconds for each of eight octave bands should be generated. Octave band–limited noise is played into the space through a low directivity loudspeaker. The noise is gated on for one second and off for 1 second. The room decay is evaluated during the off span. If it decays completely before the next burst, the T_{30} is less than one second. If not, the next burst should be on for 2 seconds and off for 2 seconds. The measurer simply keeps advancing to the next track until the room completely decays in the off span, Figs. 46-6, 46-7, and 46-8. The advantages of this method include:

1. No sophisticated instrumentation is required.
2. The measurer is free to wander the space.
3. The nature of the decaying field can be judged.
4. A group can perform the measurement.

A test of this type is useful as a prelude to more sophisticated techniques.

46.3.2.2. Amplitude versus Time

Fig. 46-9 shows an audio waveform displayed as amplitude versus time. This representation is especially meaningful to humans since it can represent the motion of the eardrum about its resting position. The waveform shown is of a male talker recorded in an anechoic (echo-free) environment. The 0 line represents the ambient (no signal) state of the medium being modulated. This would be ambient atmospheric pressure for an acoustical wave, or zero volts or a dc offset for an electrical waveform measured at the output of a system component.

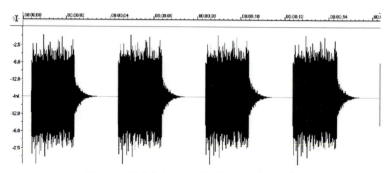

Figure 46-6 A room with $RT_{60} < 2$ seconds.

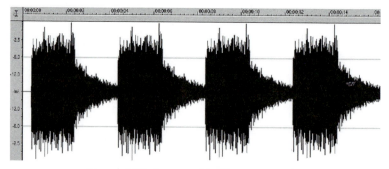

Figure 46-7 A room with $RT_{60} > 2$ seconds.

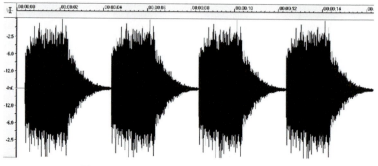

Figure 46-8 A room with $RT_{60} = 2$ seconds.

Fig. 46-10 shows the same waveform, but this time played over a loud-speaker into a room and recorded. The waveform has now been encoded (convolved) with the response of the loudspeaker and room. It will sound completely different than the anechoic version.

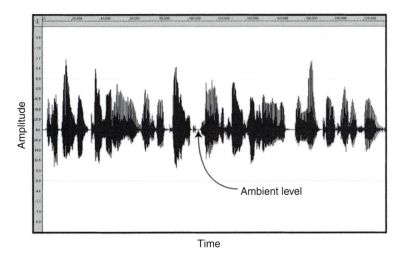

Time

Figure 46-9 Amplitude versus time plot of a male talker made in an anechoic environment.

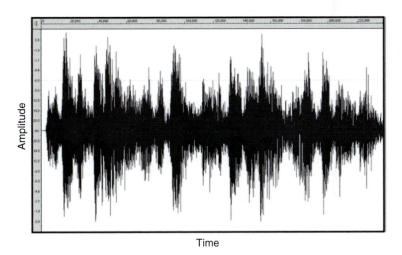

Time

Figure 46-10 The voice waveform after encoding with the room response.

Fig. 46-11 shows an impulse response and Fig. 46-12 shows the envelope-time curve (ETC) of the loudspeaker and room. It is essentially the difference between Fig. 46-9 and Fig. 46-10 that fully characterizes any effect that the loudspeaker or room has on the electrical signal fed to the loudspeaker and measured at that point in space. Most measurement systems attempt to measure the impulse response, since knowledge of the impulse

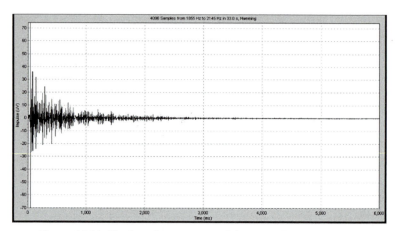

Figure 46-11 The impulse response of the acoustic environment.

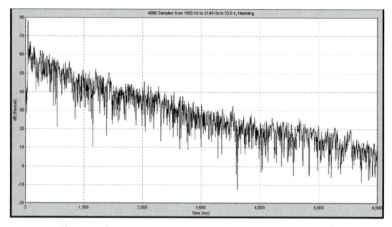

Figure 46-12 The envelope-time curve (ETC) of the same environment. It can be derived from the impulse response.

response of a system allows its effect on any signal passing through it to be determined, assuming the system is linear and time invariant. This effect is called the transfer function of the system and includes both magnitude (level) and phase (timing) information for each frequency in the pass band. Both the loudspeaker and room can be considered filters that the energy must pass through en route to the listener. Treating them as filters allows their responses to be measured and displayed, and provides an objective benchmark to evaluate their effect. It also opens loudspeakers and rooms to evaluation by electrical network analysis methods, which are generally more widely known and better developed than acoustical measurement methods.

46.3.2.3. The Transfer Function

The effect that a filter has on a waveform is called its *transfer function*. A transfer function can be found by comparing the input signal and output signal of the filter. It matters little if the filter is an electronic component, loudspeaker, room, or listener. The time domain behavior of a system (impulse response) can be displayed in the frequency domain as a spectrum and phase (transfer function). Either the time or frequency description fully describes the filter. Knowledge of one allows the determination of the other. The mathematical map between the two representations is called a transform. Transforms can be performed at amazingly fast speeds by computers. Fig. 46-13 shows a domain chart that provides a map between various representations of a system's response. The measurer must remember that the responses being measured and displayed on the analyzer are dependent on the test stimulus used to acquire the response. Appropriate stimuli must have adequate energy content over the pass band of the system being measured. In other words, we can't measure a subwoofer using a flute

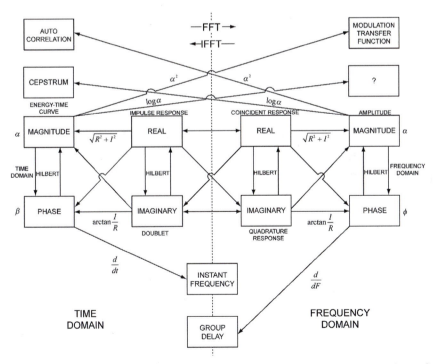

Figure 46-13 The domain chart provides a map between various representations of a system response. Courtesy Brüel and Kjaer.

solo as a stimulus. With that criteria met, the response measured and displayed on the analyzer is independent of the program material that passes through a linear system. Pink noise and sine sweeps are common stimuli due to their broadband spectral content. In other words, the response of the system doesn't change relative to the nature of the program material. For a linear system, the transfer function is a summary that says, "If you put energy into this system, this is what will happen to it."

The domain chart provides a map between various methods of displaying the system's response. The utility of this is that it allows measurement in either the time or frequency domain. The alternate view can be determined mathematically by use of a transform. This allows frequency information to be determined with a time domain measurement, and time information to be determined by a frequency domain measurement. This important inverse relationship between time and frequency can be exploited to yield many possible ways of measuring a system and/or displaying its response. For instance, a noise immunity characteristic not attainable in the time domain may be attainable in the frequency domain. This information can then be viewed in the time domain by use of a transform. The Fourier Transform and its inverse are commonly employed for this purpose. Measurement programs like Arta can display the signal in either domain, Fig. 46-14.

46.3.3. Measurement Systems

Any useful measurement system must be able to extract the system response in the presence of noise. In some applications, the signal-to-noise requirements might actually determine the type of analysis that will be used. Some of the simplest and most convenient tests have poor signal-to-noise performance, while some of the most complex and computationally demanding methods can measure under almost any conditions. The measurer must choose the type of analysis with these factors in mind. It is possible to acquire the impulse response of a filter without using an impulse. This is accomplished by feeding a known broadband stimulus into the filter and reacquiring it at the output. A complex comparison of the two signals (mathematical division) yields the transfer function, which is displayed in the frequency domain as a magnitude and phase or inverse-transformed for display in the time domain as an impulse response. The impulse response of a system answers the question, "If I feed a perfect impulse into this system, when will the energy exit the system?" A knowledge of "when" can characterize

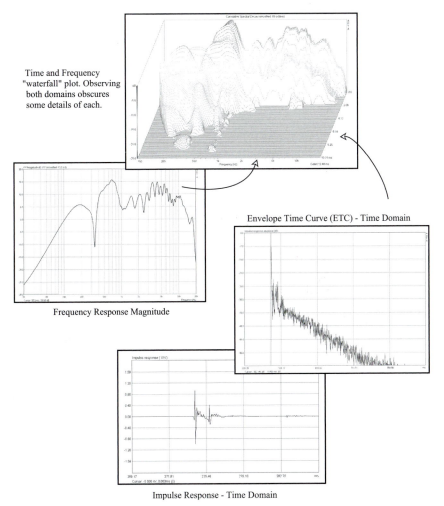

Time and Frequency "waterfall" plot. Observing both domains obscures some details of each.

Envelope Time Curve (ETC) - Time Domain

Frequency Response Magnitude

Impulse Response - Time Domain

Figure 46-14 The FFT can be used to view the spectral content of a time domain measurement, Arta 1.2.

a system. After transformation, the spectrum or frequency response is displayed on a decibel scale. A phase plot shows the phase response of the device-under-test, and any phase shift versus frequency becomes apparent. If an impulse response is a measure of when, we might describe a frequency response as a measure of what. In other words, "If I input a broadband stimulus (all frequencies) into the system, what frequencies will be present at the output of the system and what will their phase relationship be?" A transfer function includes both magnitude and phase information.

46.3.3.1. Alternate Perspectives

The time and frequency views of a system's response are mutually exclusive. By definition the time period of a periodic event is

$$T = \frac{1}{f}$$ 46-1

where,
T is time in seconds,
f is frequency in hertz.

Since time and frequency are reciprocals, a view of one excludes a view of the other. Frequency information cannot be observed on an impulse response plot, and time information can't be observed on a magnitude/phase plot. Any attempt to view both simultaneously will obscure some of the detail of both. Modern analyzers allow the measurer to switch between the time and frequency perspectives to extract information from the data.

46.3.4. Testing Methods

Compared to other components in the sound system, the basic design of loudspeakers and compression drivers has changed relatively little in the last 50 years. At over a half-century since their invention, we are still pushing air with pistons driven by voice coils suspended in magnetic fields. But the methods for measuring their performance have improved steadily since computers can now efficiently perform digital sampling and signal processing, and execute transforms in fractions of a second. Extremely capable measurement systems are now accessible and affordable to even the smallest manufacturers and individual audio practitioners. A common attribute of systems suitable for loudspeaker testing is the ability to make reflection-free measurements indoors, without the need for an anechoic chamber. Anechoic measurements in live spaces can be accomplished by the use of a time window that allows the analyzer to collect the direct field response of the loudspeaker while ignoring room reflections. Conceptually, a time window can be thought of as an accurate switch that can be closed as the desired waves pass the microphone and opened prior to the arrival of undesirable reflections from the environment. A number of implementations exist, each with its own set of advantages and drawbacks. The potential buyer must understand the trade-offs and choose a system that offers the best set of compromises for the intended application. Parameters of interest include signal-to-noise ratios, speed, resolution, and price.

46.3.4.1. FFT Measurements

The Fourier Transform is a mathematical filtering process that determines the spectral content of a time domain signal. The Fast Fourier Transform, or FFT, is a computationally efficient version of the same. Most modern measurement systems make use of the computer's ability to quickly perform the FFT on sampled data. The cousin to the FFT is the IFFT, or Inverse Fast Fourier Transform. As one might guess, the IFFT takes a frequency domain signal as its input and produces a time domain signal. The FFT and IFFT form the bedrock of modern measurement systems. Many fields outside of audio use the FFT to analyze time records for periodic activity, such as utility companies to find peak usage times or an investment firm to investigate cyclic stock market behavior. Analyzers that use the Fast Fourier Transform to determine the spectral content of a time-varying signal are collectively called FFTs. If a broadband stimulus is used, the FFT can show the spectral response of the device under test (DUT). One such stimulus is the unit impulse, a signal of theoretically infinite amplitude and infinitely small time duration. The FFT of such a stimulus is a straight, horizontal line in the frequency domain.

The time-honored hand clap test of a room is a crude but useful form of impulse response. The hand clap is useful for casual observations, but more accurate and repeatable methods are usually required for serious audio work. The drawbacks of using impulsive stimuli to measure a sound system include:

1. Impulses can drive loudspeakers into nonlinear behavior.
2. Impulse responses have poor signal-to-noise ratios, since all of the energy enters the system at one time and is reacquired over a longer span of time along with the noise from the environment.
3. There is no way to create a perfect impulse, so there will always be some uncertainty as to whether the response characteristic is that of the system, the impulse, or some nonlinearity arising from impulsing a loudspeaker.

Even with its drawbacks, impulse testing can provide useful information about the response of a loudspeaker or room.

46.3.4.2. Dual-Channel FFT

When used for acoustic measurements, dual-channel FFT analyzers digitally sample the signal fed to the loudspeaker, and also digitally sample the acoustic signal from the loudspeaker at the output of a test microphone. The signals are then compared by division, yielding the transfer function of the loudspeaker. Dual-channel FFTs have the advantage of being able to use any broadband stimulus as a test signal. This advantage is offset somewhat by poorer signal-to-noise performance and stability than other types of measurement systems, but the performance is often adequate for many measurement chores. Pink noise and swept sines provide much better stability and noise immunity. It is a computationally intense method since both the input and output signal must be measured simultaneously and compared, often in real time. For a proper comparison to yield a loudspeaker transfer function, it is important that the signals being compared have the same level, and that any time offsets between the two signals be removed. Dual-channel FFT analyzers have set up routines that simplify the establishment of these conditions. Portable computers have A/D converters as part of their on-board sound system, as well as a microprocessor to perform the FFT. With the appropriate software and sound system interface they form a powerful, low-cost and portable measurement platform.

46.3.4.3. Maximum-Length Sequence

The maximum-length sequence (MLS) is a pseudorandom noise test stimulus. The MLS overcomes some of the shortcomings of the dual-channel FFT, since it does not require that the input signal to the system be measured. A binary string (ones and zeros) is fed to the device under test while simultaneously being stored for future correlation with the loud-speaker response acquired by the test microphone. The pseudorandom sequence has a white spectrum (equal energy per Hz), and is exactly known and exactly repeatable. Comparing the input string with the string acquired by the test microphone yields the transfer function of the system. The advantage of the MLS is its excellent noise immunity and fast measurement time, making it a favorite of loudspeaker designers. A disadvantage is that the noiselike stimulus can be annoying, sometimes requiring that measurements be done after hours. The use of MLS has waned in recent years to log-swept sine measurements made on dual-channel FFT analyzers.

46.3.4.4. Time-Delay Spectrometry (TDS)

TDS is a fundamentally different method of measuring the transfer function of a system. Richard Heyser, a staff scientist at the Jet Propulsion Laboratories, invented the method. An anthology of Mr. Heyser's papers on TDS is available in the reference. Both the dual-channel FFT and MLS methods involve digital sampling of a broadband stimulus. TDS uses a method borrowed from the world of sonar, where a single-frequency sinusoidal "chirp" signal is fed to the system under test. The chirp slowly sweeps through the frequencies being measured, and is reacquired with a tracking filter by the TDS analyzer. The reacquired signal is then mixed with the outgoing signal, producing a series of sum and difference frequencies, each frequency corresponding to a different arrival time of sound at the microphone. The difference frequencies are transformed to the time domain with the appropriate transform, yielding the envelope-time Curve (ETC) of the system under test. TDS is based on the frequency domain, allowing the tracking filter to be tuned to the desired signal while ignoring signals outside of its bandwidth. TDS offers excellent noise immunity, allowing good data to be collected under near-impossible measurement conditions. Its downside is that good low-frequency resolution can be difficult to obtain without extended measurement times, plus the correct selection of measurement parameters requires a knowledgeable user. In spite of this, it is a favorite among contractors and consultants, who must often perform sound system calibrations in the real world of air conditioners, vacuum cleaners, and building occupants.

While other measurement methods exist, the ones outlined above make up the majority of methods used for field and lab testing of loudspeakers and rooms. Used properly, any of the methods can provide accurate and repeatable measured data. Many audio professionals have several measurement platforms and exploit the strong points of each when measuring a sound system.

46.3.5. Preparation

There are many measurements that can be performed on a sound system. A prerequisite to any measurement is to answer the following questions:
1. What am I trying to measure?
2. Why am I trying to measure it?
3. Is it audible?
4. Is it relevant?

Failure to consider these questions can lead to hours of wasted time and a hard drive full of meaningless data. Even with the incredible technologies that we have available to us, the first part of any measurement session is to listen. It can take many hours to determine what needs to be measured to solve a sound system problem, yet the actual measurement itself can often be completed in seconds. Using an analogy from the medical field, the physician must query the patient at length to narrow down the ailment. The more that is known about the ailment, the more specific and relevant the tests that can be run for diagnosis. There is no need to test for tonsillitis if the problem is a sore back!

1. What am I measuring? A fundamental decision that precedes a meaningful measurement is how much of the room's response to include in the measured data. Modern measurement systems have the ability to perform semianechoic measurements, and the measurer must decide if the loudspeaker, the room, or the combination needs to be measured. If one is diagnosing loudspeaker ailments, there is little reason to select a time window long enough to include the effects of late reflections and reverberation. A properly selected time window can isolate the direct field of the loudspeaker and allow its response to be evaluated independently of the room. If one is trying to measure the total decay time of the room, the direct sound field becomes less important, and a microphone placement and time window are selected to capture the entire energy decay. Most modern measurement systems acquire the complete impulse response, including the room decay, so the choice of the time window size can be made after the fact during post processing.

2. Why am I measuring? There are several reasons for performing acoustic measurements in a space. An important reason for the system designer is to characterize the listening environment. Is it dead? Is it live? Is it reverberant? These questions must be considered prior to the design of a sound system for the space. While the human hearing system can provide the answers to these questions, it cannot document them and it is easily deceived. Measurements might also be performed to document the performance of an existing system prior to performing changes or adding room treatment. Customers sometimes forget how bad it once sounded after a new or upgraded system is in place for a few weeks.

   The most common reason for performing measurements on a system is for calibration purposes. This can include equalization, signal alignment, crossover selection, and a multiplicity of other reasons. Since loudspeakers interact in a complex way with their environment, the final phase of any system installation is to verify system performance by measurement.

3. Is it audible? Can I hear what I am trying to measure? If one cannot hear an anomaly, there is little reason to attempt to measure it. The human hearing system is perhaps the best tool available for determining what should be measured about a sound system. The human hearing system can tell us that something doesn't sound right, but the cause of the

problem can be revealed by measurement. Anything you can hear can be measured, and once it is measured it can be quantified and manipulated.

4. Is it relevant? Am I measuring something that is worth measuring? If one is working for a client, time is money. Measurements must be prioritized to focus on audible problems. Endless hours can be spent "chasing rabbits" by measuring details that are of no importance to the client. This is not necessarily a fruitless process, but it is one that should be done on your own time. I have on several occasions spent time measuring and documenting anomalies that had nothing to do with the customer's reason for calling me. All venues have problems that the owner is unaware of. Communication with the client is the best way to avoid this pitfall.

46.3.5.1. Dissecting the Impulse Response

The audio practitioner is often faced with the dilemma of determining whether the reason for bad sound is the loudspeaker system, the room, or an interaction of the two. The impulse response can hold that answer to these and other perplexing questions. The impulse response in its amplitude versus time display is not particularly useful for other than determining the polarity of a system component, Fig. 46-15. A better representation comes from squaring impulse response (making all deflections positive) and displaying the square root of the result on a logarithmic vertical scale. This

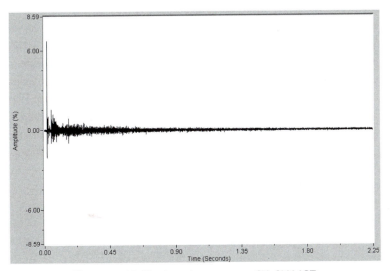

Figure 46-15 The impulse response, SIA-SMAART.

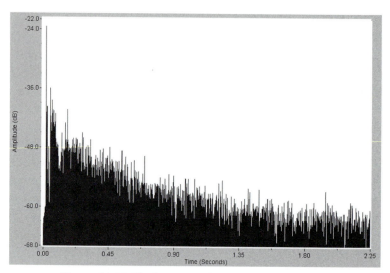

Figure 46-16 The log-squared response, SIA-SMAART.

log-squared response allows the relative levels of energy arrivals to be compared, Fig. 46-16.

46.3.5.2. The Envelope-Time Curve

Another useful way of viewing the impulse response is in the form of the envelope-time curve, or ETC. The ETC is also a contribution of Richard Heyser.[2] It takes the real part of the impulse response and combines it with a 90 degrees phase shifted version of the same, Fig. 46-17. One way to get the shifted version is to use the Hilbert Transform. The complex combination of these two signals yields a time domain waveform that is often easier to interpret than the impulse response. The ETC can be loosely thought of as a smoothing function for the log-squared response, showing the envelope of the data. This can be more revealing as to the audibility of an event. The impulse response, log-squared response, and energy-time curve are all different ways to view the time domain data.

46.3.5.3. A Global Look

When starting a measurement session, a practical approach is to first take a global look and measure the complete decay of the room. The measurer can then choose to ignore part of the time record by using a time window to isolate the desired part during post processing. The length of the time window can be increased to include the effects of more of the energy

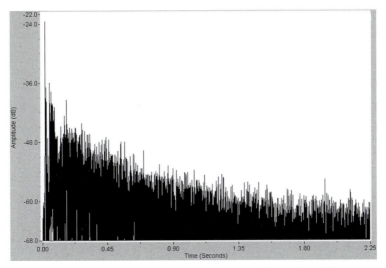

Figure 46-17 The envelope-time curve (ETC), SIA-SMAART.

returned by the room. The time window can also be used to isolate a reflection and view its spectral content. Just like your life span represents a time window in human history, a time window can be used to isolate parts of the impulse response.

46.3.5.4. Time Window Length

The time domain response can be divided to identify the portion that can be attributed to the loudspeaker and that which can be attributed to the room. It must be emphasized that there is a rather gray and frequency-dependent line between the two, but for this discussion we will assume that we can clearly separate them. The direct field is the energy that arrives at the listener prior to any reflections from the room. The division is fairly distinct if neither the loudspeaker nor microphone is placed near any reflecting surfaces, which, by the way, is a good system design practice. At long wavelengths (low frequencies) the direct field may include the effects of boundaries near the loudspeaker and microphone. As frequency increases, the sound from the loudspeaker becomes less affected by boundary effects (due in part to increased directivity) and can be measured independently of them. Proper loudspeaker placement produces a time gap between the sound energy arrivals from the loudspeaker and the later arriving room response. We can use this time gap to aid in selecting a time window to separate the loudspeaker response from the room response and diagnosing system problems.

46.3.5.5. Acoustic Wavelengths

Sound travels in waves. The sound waves that we are interested in characterizing have a physical size. There will be a minimum time span required to observe the spectral response of a waveform. The minimum required length of time to view an acoustical event is determined by the longest wavelength (lowest frequency) present in the event. At the upper limits of human hearing, the wavelengths are only a few millimeters in length, but as frequency decreases the waves become increasingly larger. At the lowest frequencies that humans hear, the wavelengths are many meters long, and can actually be larger than the listening (or measurement) space. This makes it difficult to measure low frequencies from a loudspeaker independently of the listening space, since low frequencies radiated from a loudspeaker interact (couple) with the surfaces around them. In an ideally positioned loudspeaker, the first energy arrival from the loudspeaker at mid- and high frequencies has already dissipated prior to the arrival of reflections and can therefore often be measured independently of them. The human hearing system tends to fuse the direct sound from the loudspeaker with the early reflections from nearby surfaces with regard to level (loudness) and frequency (tone). It is usually useful to consider them as separate events, especially since the time offset between the direct sound and first reflections will be unique for each listening position. This precludes any type of frequency domain correction (i.e., equalization) of the room/loudspeaker response other than at frequencies where coupling occurs due to close proximity to nearby surfaces. While it is possible to compensate to some extent for room reflections at a point in space (acoustic echo cancellers used for conference systems), this correction cannot be extended to include an area. This inability to compensate for the reflected energy at mid/high frequencies suggests that their effects be removed from the loudspeaker's direct field response prior to meaningful equalization work by use of an appropriate time window.

46.3.5.6. Microphone Placement

A microphone is needed to acquire the sound radiated into the space from the loudspeaker at a discrete position. Proper microphone placement is determined by the type of test being performed. If one were interested in measuring the decay time of the room, it is usually best to place the microphone well beyond critical distance. This allows the build-up of the reverberant field to be observed as well as providing good resolution of the decaying tail. Critical distance is the distance from the loudspeaker at which the direct field level and reverberant field level are equal. It is described further in Section 46.3.5.7. If it's the loudspeaker's response that needs to be measured, then a microphone placement inside of critical distance will provide better data on some types of analyzers, since the direct sound field is stronger relative to the later energy returning from the room. If the microphone is placed too close to the loudspeaker, the measured sound levels will be accurate for that position, but may not accurately extrapolate to greater distances with the inverse-square law. As the sound travels farther, the response at a remote listening position may bear little resemblance to the response at the near field microphone position. For this reason, it is usually desirable to place the microphone in the far free field of the loudspeaker—not too close and not too far away. The approximate extent of the near field can be determined by considering that the path length difference from the measurement position (assumed axial) and the edge of the sound radiator should be less than ¼ wavelength at the frequency of interest. This condition is easily met for a small loudspeaker that is radiating low frequencies. Such devices closely approximate an ideal point source. As the frequency increases the condition becomes more difficult to satisfy, especially if the size of the radiator also increases. Large radiators (or groups of radiators) emitting high frequencies can extend the near field to very long distances. Line arrays make use of this principle to overcome the inverse-square law. In practice, small bookshelf loudspeakers can be accurately measured at a few meters. About 10 m is a common measurement distance for moderate-sized, full-range loudspeakers in a large space. Even greater distances are required for large devices radiating high frequencies. A general guideline is to not put the mic closer than three times the loudspeaker's longest dimension.

46.3.5.7. Estimate the Critical Distance DC

Critical distance is easy to estimate. A quick method with adequate accuracy requires a sound level meter and noise source. Ideally, the noise source should be band limited, as critical distance is frequency dependent. The 2 kHz octave band is a good place to start when measuring critical distance. Proceed as follows:

1. Energize the room with pink noise in the desired octave band from the sound source being measured. The level should be at least 25 dB higher than the background noise in the same octave band.
2. Using the sound level meter, take a reading near the loudspeaker (about 1 m) and on-axis. At this distance, the direct sound field will dominate the measurement.

3. Move away from the loudspeaker while observing the sound level meter. The sound level will fall off as you move farther away. If you are in a room with a reverberant sound field, at some distance the meter reading will quit dropping. You have now moved beyond critical distance. Measurements of the direct field beyond this point will be a challenge for some types of analysis. Move back toward the loudspeaker until the meter begins to rise again. You are now entering a good region to perform acoustic measurements on loudspeakers in this environment. The above process provides an estimate that is adequate for positioning a measurement microphone for loudspeaker testing. With a mic placement inside of critical distance, the direct field is a more dominant feature on the impulse response and a time window will be more effective in removing room reflections.

At this point it is interesting to wander around the room with the sound level meter and evaluate the uniformity of the reverberant field. Rooms that are reverberant by the classical definition will vary little in sound level beyond critical distance when energized with a continuous noise spectrum. Such spaces have low internal sound absorption relative to their volume.

46.3.5.8. Common Factors to All Measurement Systems

Let's assume that we wish to measure the impulse response of a loudspeaker/room combination. While it would not be practical to measure the response at every seat, it is good measurement practice to measure at as many seats as are required to prove the performance of the system. Once the impulse response is properly acquired, any number of post processes can be performed on the data to extract information from it. Most modern measurement systems make use of digital sampling in acquiring the response of the system. The fundamentals and prerequisites are not unlike the techniques used to make any digital recording, where one must be concerned with the level of an event and its time length. Some setup is required and some fundamentals are as follows:

1. The sampling rate must be fast enough to capture the highest frequency component of interest. This requires at least two samples of the highest frequency component. If one wished to measure to 20 kHz, the required sample rate would need to be at least 40 kHz. Most measurement systems sample at 44.1 kHz or 48 kHz, more than sufficient for acoustic measurements.

2. The time length of the measurement must be long enough to allow the decaying energy curve to flatten out into the room noise floor. Care

must be taken to not cut off the decaying energy, as this will result in artifacts in the data, like a scratch on a phonograph record. If the sampling rate is 44.1 kHz, then 44,100 samples must be collected for each second of room decay. A 3-second room would therefore require 44.1 × 1000 × 3 or 128,000 samples. A hand clap test is a good way to estimate the decay time of the room and therefore the required number of samples to fully capture it. The time span of the measurement also determines the lowest frequency that can be resolved from the measured data, which is approximately the inverse of the measurement length. The sampling rate can be reduced to increase the sampling time to yield better low-frequency information. The trade-off is a reduction in the highest frequency that can be measured, since the condition outlined in step one may have been violated.

3. The measurement must hav a sufficient signal-tonoise ratio to allow the decaying tail to be fully observed. This often requires that the measurement be repeated a number of times and the results averaged. Using a dual-channel FFT or MLS, the improvement in SNR will be 3 dB for each doubling of the number of averages. Ten averages is a good place to start, and this number can be increased or decreased depending on the environment. The level of the test stimulus is also important. Higher levels produce improved SNR, but can also stress the loudspeaker.

4. Perform the test and observe the data. It should fill the screen from top left to bottom right and be fully decayed prior to reaching the right side of the screen. It should also be repeatable. Run the test several times to check for consistency. Background noise can dramatically affect the repeatability of the measurement and the validity of the data.

Once the impulse response is acquired, it can be further analyzed for spectral content, intelligibility information, decay time, etc. These are referred to as metrics, and some require some knowledge on the part of the measurer in properly placing markers (called *cursors*) to identify the parameters required to perform the calculations. Let us look at how the response of the loudspeaker might be extracted from the data just gathered.

The time domain data displays what would have resulted if an impulse were fed through the system. Don't try to correlate what you see on the analyzer with what you heard during the test. Most measurement systems display an impulse response that is calculated from a knowledge of the input and output signal to the system, and there is no resemblance between what you hear when the test is run and what you are seeing on the screen, Fig. 46-18.

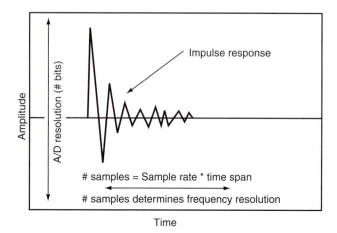

Figure 46-18 Many analyzers acquire the room response by digital sampling.

We can usually assume that the first energy arrival is from the loudspeaker itself, since any reflection would have to arrive later than the first wave front since it had to travel farther. Pre-arrivals can be caused by the acoustic wave propagating through a solid object, such as a ceiling or floor and reradiating near the microphone. Such arrivals are very rare and usually quite low in level. In some cases a reflection may actually be louder than the direct arrival. This could be due to loudspeaker design or its placement relative to the mic location. It's up to the measurer to determine if this is normal for a given loudspeaker position/seating position. All loudspeakers will have some internal and external reflections that will arrive just after the first wave front. These are actually a part of the loudspeaker's response and can't be separated from the first wave front with a time window due to their close proximity without extreme compromises in frequency resolution. Such reflections are at least partially responsible for the characteristic sound of a loudspeaker. Studio monitor designers and studio control room designers go to great lengths to reduce the level of such reflections, yielding more accurate sound reproduction. Good system design practice is to place loudspeakers as far as possible from boundaries (at least at mid- and high frequencies). This will produce an initial time gap between the loudspeaker's response and the first reflections from the room. This gap is a good initial dividing point between the loudspeaker's response and the room's response, with the energy to the left of the dividing cursor being the response of the loudspeaker and the energy to the right the response of the room. The placement of this divider can form a time window by having the analyzer ignore everything later in time than the cursor setting. The time window size also determines the frequency resolution of the postprocessed data. In the frequency domain, improved resolution means a smaller number. For instance, 10 Hz resolution is better than 40 Hz resolution. Since time and frequency have an inverse relationship, the time window length required to observe 10 Hz will be much longer than the time window length required to resolve 40 Hz. The resolution can be estimated by $f = 1/T$, where T is the length of the time window in seconds. Since a frequency magnitude plot is made up of a number of data points connected by a line, another way to view the frequency resolution is that it is the number of Hz between the data points in a frequency domain display.

The method of determination of the time window length varies with different analyzers. Some allow a cursor to be placed anywhere on the data record, and the placement determines the frequency resolution of the spectrum determined by the window length. Others require that the measurer select the number of samples to be used to form the time window, which in turn determines the frequency resolution of the time window. The window can then be positioned at different places on the time domain plot to observe the spectral content of the energy within the window, Figs. 46-19, 46-20, and 46-21.

For instance, a 1 second total time (44,100 samples) could be divided into about twenty two time windows of 2048 samples each (about 45 ms).

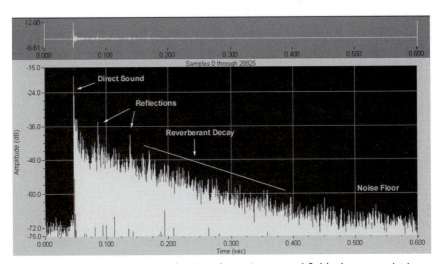

Figure 46-19 A room response showing the various sound fields that can exist in an enclosed space, SIA-SMAART.

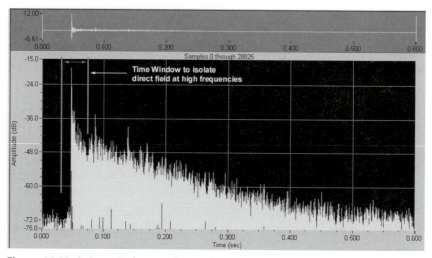

Figure 46-20 A time window can be used to isolate the loudspeaker's response from the room reflections.

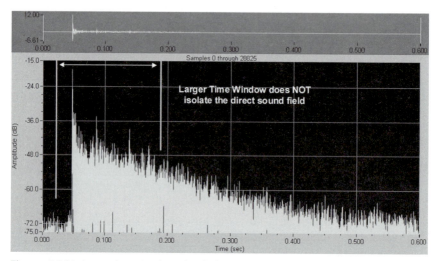

Figure 46-21 Increasing the length of the time window increases the frequency resolution, but lets more of the room into the measurement, SIA-SMAART.

Each window would allow the observation of the spectral content down to ($\frac{1}{45}$) × 1000 or 22 Hz. The windows can be overlapped and moved around to allow more precise selection of the time span to be observed. Displaying a number of these time windows in succession, each separated by a time offset, can form a 3D plot known as a waterfall.

46.3.5.9. Data Windows

There are some conditions that must be observed when placing cursors to define the time window. Ideally, we would like to place the cursor at a point on the time record where the energy is zero. A cursor placement that cuts off an energy arrival will produce a sharp rise or fall time that produces artifacts in the resultant calculated spectral response. Discontinuities in the time domain have broad spectral content in the frequency domain. A good example is a scratch on a phonograph record. The discontinuity formed by the scratch manifests itself as a broadband click during playback. If an otherwise smooth wheel has a discontinuity at one point, it would thump annoyingly when it was rolled on a smooth surface. Our measurement systems treat the data within the selected window as a continuously repeating event. The end of the event must line up with the beginning or a discontinuity occurs resulting in the generation of high-frequency artifacts called *spectral leakage*. In the same manner that a physical discontinuity in a phonograph record or wheel can be corrected by polishing, a discontinuity in a sampled time measurement can be remedied by tapering the energy at the beginning and end of the window to zero using a mathematical function. A number of data window shapes are available for performing the smoothing.

These include the Hann, Hamming, Black man-Harris, and others. In the same way that a physical polishing process removes some good material from what is being rubbed, data windows remove some good data in the process of smoothing the discontinuity. Each window has a particular shape that leaves the data largely untouched at the center of the window but tapers it to varying degrees toward the edges. Half windows only smooth the data at the right edge of the time record while full windows taper both (start and stop) edges. Since all windows have side effects, there is no clear preference as to which one should be used. The Hann window provides a good compromise between time record truncation and data preservation. Figs. 46-22 and 46-23 show how a data window might be used to reduce spectral leakage.

46.3.5.10. A Methodical Approach

Since there are an innumerable number of tests that can be performed on a system, it makes sense to establish a methodical and logical process for the measurement session. One such scenario may be as follows:

1. Determine the reason for and scope of the measurement session. What are you looking for? Can you hear it? Is it repeatable? Why do you need this information?

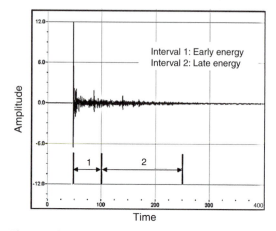

Figure 46-22 The impulse response showing both early and late energy arrivals.

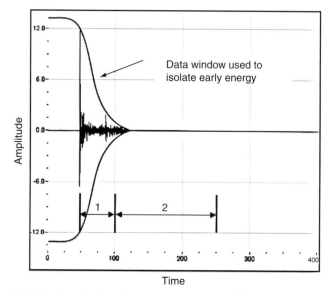

Figure 46-23 A data window is used to remove the effects of the later arrivals.

2. Determine what you are going to measure. Are you looking at the room or at the sound system? If it is the room, possibly the only meaningful measurements will be the overall decay time and the noise floor. If you are looking at the sound system, decide if you need to switch off or disconnect some loudspeakers. This may be essential to determine whether the individual components are working properly, or that an

anomaly is the result of interaction between several components. "Divide and conquer" is the axiom.

3. Select the microphone position. I usually begin by looking at the on-axis response of the loudspeaker as measured from inside of critical distance. If multiple loudspeakers are on, turn all of them off but one prior to measuring. The microphone should be placed in the far free field of the loudspeaker as previously described. When measuring a loudspeaker's response, care should be taken to eliminate the effects of early reflections on the measured data, as these will generate acoustic comb filters that can mask the true response of the loudspeaker. In most cases the predominant offending surface will be the floor or other boundaries near the microphone and loudspeaker. These reflections can be reduced or eliminated by using a ground plane microphone placement, a tall microphone stand (when the loudspeaker is overhead), or some strategically placed absorption. I prefer the tall microphone stand for measuring installed systems with seating present since it works most anywhere, regardless of the seating type. The idea is to intercept the sound on its way to a listener position, but before it can interact with the physical boundaries around that position. These will always be unique to that particular seat, so it is better to look at the free field response, as it is the common denominator to many listener seats.

4. Begin with the big picture. Measure an impulse response of the complete decay of the space. This yields an idea of the overall properties of the room/system and provides a good point of reference for zooming in to smaller time windows. Save this information for documentation purposes, as later you may wish to reopen the file for further processing.

5. Reduce the size of the time window to eliminate room reflections. Remember that you are trading off frequency resolution when truncating the time record, Fig. 46-24. Be certain to maintain sufficient resolution to allow adequate low-frequency detail. In some cases, it may be impossible to maintain a sufficiently long window to view low frequencies and at the same time eliminate the effects of reflections at higher frequencies, Fig. 46-25. In such cases, the investigator may wish to use a short window for looking at the high-frequency direct field, but a longer window for evaluating the woofer. Windows appropriate for each part of the spectrum can be used. Some measurement systems provide variable time windows, which allow low frequencies to be

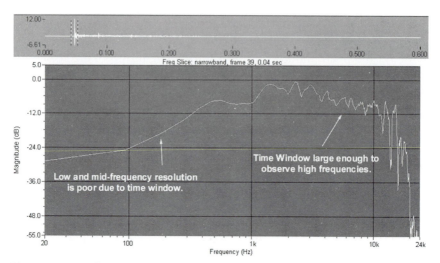

Figure 46-24 A short time window isolates the direct field at high frequencies at the expense of low-frequency resolution, SIA-SMAART.

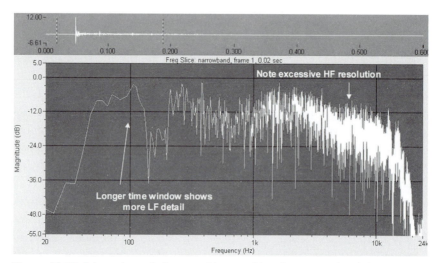

Figure 46-25 A long time window provides good low-frequency detail, SIA-SMAART.

viewed in great detail (long time window) while still providing a semi-anechoic view (short time window) at high frequencies. There is evidence to support that this is how humans process sound information, making this method particularly interesting, Fig. 46-26.

6. Are other microphone positions necessary to characterize this loud-speaker? The off-axis response of some loudspeakers is very similar to the

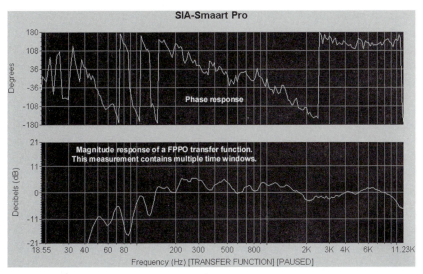

Figure 46-26 The time window increases with length as frequency decreases.

on-axis response, reducing the need to measure at many angles. Other loudspeakers have very erratic responses, and a measurement at any one point around the loudspeaker may bear little resemblance to the response at other positions. This is a design issue, but one that must be considered by the measurer.

7. Once an accurate impulse response is measured, it can be postprocessed to yield information on spectral content, speech intelligibility, and music clarity. There are a number of metrics that can provide this information. These are interpretations of the measured data and generally correlate with subjective perception of the sound at that seat.

8. An often overlooked method of evaluating the impulse response is the use of convolution to encode it onto anechoic program material. An excellent freeware convolver called GratisVolver is available from www. catt.se. Listening to the IR can often reveal subtleties missed by the various metrics, as well as provide clues as to what post process must be used to observe the event of interest.

46.3.6. Human Perception

Useful measurement systems can measure the impulse response of a loud-speaker/room combination with great detail. Information regarding speech intelligibility and music clarity can be derived from the impulse response. In nearly all cases, this involves post processing the impulse response using one of several clarity measure metrics.

46.3.6.1. Percentage Articulation Loss of Consonants-(%Alcons)

For speech, one such metric is the percentage articulation loss of conso-nants, or *%Alcons*. Though not in widespread use today, a look at it can provide insight into the requirements for good speech intelligibility. A *%Alcons* measurement begins with an impulse response, which is usually displayed as a log-squared response or ETC. Since the calculation essentially examines the ratio between early energy, late energy, and noise, the measurer must place cursors on the display to define these parameters. These cursors may be placed automatically by the measurement program. The result is weighted with regard to decay time, so this too must be defined by the measurer. Analyzers such as the TEF25™ and EASERA include best guess default placements based on the research of Peutz, Davis, and others, Fig. 46-27.

These placements were determined by correlating measured data with live listener scores in various acoustic environments, and represent a defined and orderly approach to achieving meaningful results that correlate with the perception of live listeners. The measurer is free to choose alternate cursor placements, but great care must be taken to be consistent. Also, alternate cursor placements make it difficult if not

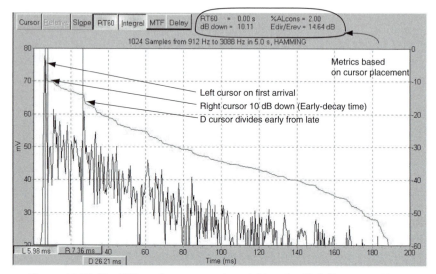

Figure 46-27 The ETC can be processed to yield an intelligibility score, TEF25.

impossible to compare your results with those obtained by other measurers. In the default %Alcons placement, the early energy (direct sound field) includes the first major sound arrival and any energy arrivals within the next 7–10 ms. This forms a tight time span for the direct sound. Energy beyond this span is considered late energy and an impairment to communication. As one might guess, a later cursor placement yields better intelligibility scores, since more of the room response is being considered beneficial to intelligibility. As such, the default placement yields a worst-case scenario. The default placement considers the effects of the early-decay time (EDT) rather than the classical T_{30} since short EDTs can yield good intelligibility, even in rooms with a long T_{30}. Again, the measurer is free to select an alternative cursor placement for determining the decay time used in the calculation, with the same caveats as placing the early-to-late dividing cursor. The %Alcons score is displayed instantly upon cursor placement and updates as the cursors are moved.

46.3.6.2. Speech Transmission Index—(STI)

The STI can be calculated from the measured impulse response with a routine outlined by Schroeder and detailed by Becker in the reference. The STI is probably the most widely used contemporary measure of intelligibility. It is supported by virtually all measurement platforms, and some handheld analyzers are available for quick checks. In short, it is a number ranging from 0 to 1, with fair intelligibility centered at 0.5 on the scale. For more details on the Speech Transmission Index, see the chapter on speech intelligibility in this text.

46.3.7. Polarity

Good sound system installation practice dictates maintaining proper signal polarity from system input to system output. An audio signal waveform always swings above and below some reference point. In acoustics, this reference point is the ambient atmospheric pressure. In an electronic device, the reference is the 0 VA reference of the power supply (often called *signal ground*) in push–pull circuits or a fixed dc offset in class A circuits. Let's look at the acoustic situation first. An increase in the air pressure caused by a sound wave will produce an inward deflection of the diaphragm of a pressure microphone (the most common type) regardless of the microphone's orientation toward the source. This inward deflection should cause a positive-going voltage swing at the output of the microphone on pin 2 relative to pin 3, as well as at the output of each piece of equipment that the signal passes through. Ultimately the electrical signal will be applied to a loudspeaker, which should deflect outward (toward an axial listener) on the positive-going signal, producing an increase in the ambient atmospheric pressure. Think of the microphone diaphragm and loudspeaker diaphragm moving in tandem and you will have the picture. Since most sound reinforcement equipment uses bipolar power supplies (allowing the audio signal to swing positive and negative about a zero reference point), it is possible for signals to become inverted in polarity (flipped over). This causes a device to output a negative-going voltage when it is fed a positive-going voltage. If the loudspeaker is reverse-polarity from the microphone, an increase in sound pressure at the microphone (compression) will cause a decrease in pressure in front of the loudspeaker (rarefaction). Under some conditions, this can be extremely audible and destructive to sound quality. In other scenarios it can be irrelevant, but it is always good to check.

System installers should always check for proper polarity when installing the sound system. There are a number of methods, some simple and some complex. Let's deal with them in order of complexity, starting with the simplest and least-costly method.

46.3.7.1. The Battery Test

Low-frequency loudspeakers can be tested using a standard 9 V battery. The battery has a positive and negative terminal, and the spacing between the terminals is just about right to fit across the terminals of most woofers. The loudspeaker cone will move outward when the battery is placed across the loudspeaker terminals with the battery positive connected to the loudspeaker positive. While this is one of the most accurate methods for testing polarity, it doesn't work for most electronic devices or high-frequency drivers. Even so, it's probably the least-costly and most accurate way to test a woofer.

46.3.7.2. Polarity Testers

There are a number of commercially available polarity test sets in the audio marketplace. The set includes a sending device that outputs a test pulse, Fig. 46-28, through a small loudspeaker (for testing microphones) or an XLR connector (for testing electronic devices) and a receiving device that collects the signal via an internal microphone (loudspeaker testing) or XLR input jack. A green light indicates correct polarity and a red light indicates

Figure 46-28 A popular polarity test set.

reverse polarity. The receive unit should be placed at the system output (in front of the loudspeaker) while the send unit is systematically moved from device to device toward the system input. A polarity reversal will manifest itself by a red light on the receive unit.

46.3.7.3. Impulse Response Tests

The impulse response is perhaps the most fundamental of audio and acoustic measurements. The polarity of a loudspeaker or electronic device can be determined from observing its impulse response, Figs. 46–29 and 46–30. This

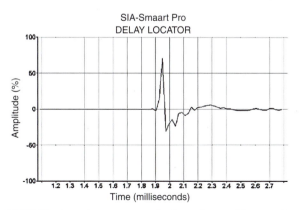

Figure 46-29 The impulse response of a transducer with correct polarity.

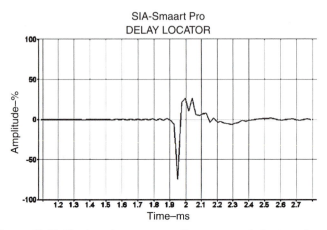

Figure 46-30 The impulse response of a reverse-polarity transducer.

is one of the few ways to test flown loudspeakers from a remote position. It is best to test the polarity of components of multiway loudspeakers individually, since all of the individual components may not be polarized the same. Filters in the signal path (i.e., active crossover network) make the results more difficult to interpret, so it may be necessary to carefully test a system component (i.e., woofer) full-range for definitive results. Be sure to return the crossover to its proper setting before continuing.

46.4. CONCLUSION

The test and measurement of the sound reinforcement system are a vital part of the installation and diagnostic processes. The FFT and the analyzers that use it have revolutionized the measurement process, allowing sound practitioners to pick apart the system response and look at the response of the loudspeaker, roo, or both. Powerful analyzers that were once beyond the reach of most technicians are readily available and affordable, and cost can no longer be used as an excuse for not measuring the system. The greatest investment by far is the time required to grasp the fundamentals of acoustics to allow interpretation of the data. Some of this information is general, and some of it is specific to certain measurement systems.

The acquisition of a measurement system is the first step in ascending the capability and credibility ladder. The next steps include acquiring proper instruction on its use by self-study or short course. The final and most important steps are the countless hours in the field required to correlate measured data with the hearing process. As proficiency in this area increases, the speed of execution, validity, and relevance of the measurements will increase also. While we can all learn how to make the measurements in a relatively short time span, the rest of our careers will be spent learning how to interpret what we are measuring.

REFERENCES

1. The Heyser Anthology, Audio Engineering Society, NY, NY.

BIBLIOGRAPHY

Davis, D., & Davis, C. (1997). *Sound System Engineering*. Boston, MA: Focal Press.
Understanding Sound System/Room Interactions, Sam Berkow, Syn-Aud-Con Tech Topic, Volume 28, Number 2.
Joseph D'Appolitto, *Testing Loudspeakers*, Old Colony Sound Labs, Petersborough, NH.